STUDENT RESOURCE BOOK

Mathematics UNLIMITED

HOLT, RINEHART and WINSTON, INC.
Austin   New York   San Diego   Chicago   Toronto   Montreal

AUTHORS

Francis "Skip" Fennell
Chairman, Education Department
Associate Professor of Education
Western Maryland College
Westminster, Maryland

Robert E. Reys
Professor of Mathematics Education
University of Missouri
Columbia, Missouri

Barbara J. Reys
Assistant Professor of Curriculum
and Instruction
University of Missouri, Columbia, Missouri
Formerly Junior High Mathematics Teacher
Oakland Junior High, Columbia, Missouri

Arnold W. Webb
Senior Research Assoiciate
Research for Better Schools
Philadelphia, Pennsylvania
Formerly Asst. Commissioner of Education
New Jersey State Education Department

CONTENTS

PRACTICE	1–170
RETEACH	1–118

Copyright © 1988 by **HOLT, RINEHART and WINSTON, INC.**

All rights reserved. No part of this publication may be reproduced or transmitted in any form or by any means, electronic or mechanical, including photocopying, recording, or any information storage and retrieval system, without permission in writing from the publisher.

Requests for permission to make copies of any part of the work should be mailed to: Permissions, Holt, Rinehart and Winston, Inc., 1627 Woodland Avenue, Austin, Texas 78741

Printed in United States of America

ISBN 0-03-021912-4

STUDENT RESOURCE BOOK

TO THE TEACHER:

The Student Resource Book contains Practice and Reteach sections that provide additional instruction, extra practice, and review of material covered in the pupil's edition. This book provides a hardbound, non-consumable resource to meet the individual needs of each student.

PRACTICE

This section provides additional practice for lessons found in the pupil's edition. Each page is keyed to the appropriate lesson, but could be used as extra practice or review at any time after the lesson.

RETEACH

This section provides additional instruction on material covered in the pupil's edition. Each page is keyed to the appropriate lesson, but could be used as instruction, extra practice, or review at any time after the lesson.

PRACTICE: Numbers to Hundred Thousands

Write the word name for each number.

1. 411,412 = _____
2. 55,132 = _____
3. 699,005 = _____

Write each number in expanded form.

4. 981 = 900 + 80 + 1 =
5. 4,198 = 4,000 + 100 + 90 + 8 =
6. 59,761 = 50,000 + 9,000 + 700 + 60 + 1 =
7. 731,428 = 700,000 + 30,000 + 1,000 + 400 + 20 + 8 =

Solve.

8. In order to fly around Earth, one must travel about twenty-four thousand, nine hundred two miles. Write the number in standard form.

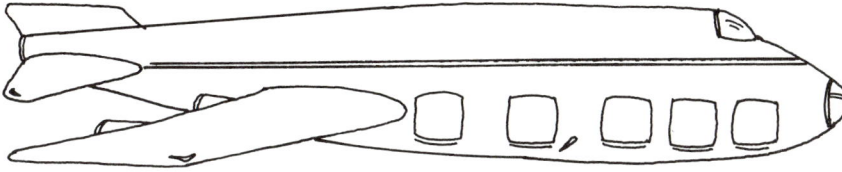

Write the number in standard form.

9. 14,000 + 300 + 20 = 14,320
10. 500,000 + 40,000 + 1,000 + 200 + 30 + 2 = 541,232
11. two thousand nine = 2,009
12. four hundred ninety-eight thousand ten = 49810

Use with pages 2–3.

PRACTICE — Numbers to Hundred Billions

PERIOD	Billions			Millions			Thousands			Ones		
	hundred billions	ten billions	billions	hundred millions	ten millions	millions	hundred thousands	ten thousands	thousands	hundreds	tens	ones
	4	6	7	5	3	2	9	7	1	6	3	8

Use the chart. Write the digits in the

1. millions period. __532__
2. ones period. __638__
3. thousands period. __971__
4. billions period. __467__

Write the number in standard form.

5. six hundred ninety-eight billion, seven hundred fifty-one million = __698751__

6. nine hundred seventy-eight million, six hundred forty-one thousand = _____

7. five hundred eighty-one billion, nine hundred sixty-two thousand = _____

For 798,631,424,562, write the digit for each place.

8. ten millions _____
9. thousands _____
10. ones _____
11. ten billions _____
12. hundred billions _____
13. ten thousands _____

Use with pages 4–5.

PRACTICE — Comparing and Ordering Numbers

Compare. Write >, <, or = for ◯.

1. 126,365 < 177,263
2. 4,788 < 4,878
3. 7,838,656 > 7,799,699
4. 653 = 653
5. 62,017 > 16,199
6. 21 < 394
7. 45,604 < 405,604
8. 10,121 > 10,101
9. 1,114 < 19,000
10. 1,657 = 1,657
11. 200 million, 777 thousand = 200 million, 777 thousand
12. 1 billion, 33 thousand > 1 billion, 30 thousand
13. 100 billion, 306 thousand > 15 billion, 306 thousand

Write in order from the least to the greatest.

14. 140,200; 148,000; 144,811; 140,147 _148,000 144,811 140,200 140,147_

15. 91,767; 99,009; 107,706; 110,421 _____

16. 320,201; 322,221; 322,201; 322,222 _____

Write in order from the greatest to the least.

17. 781,707; 423,792; 244,711 _____

18. 11,765; 14,000; 790; 51,080 _____

19. 369,029; 126,769; 41,093; 72,492 _____

Use with pages 6–7.

PRACTICE Using the Help File

Write the letter of the idea from the Help File needed to help each student solve the word problem.

In the search for minerals, water, and fuel, miners have drilled holes in the earth's surface thousands of feet deep. Many of the deepest drilling sites are in the United States. In Louisiana, depths of 22,570 ft and 25,600 ft have been reached. In California, a hole was drilled to 21,842 ft, and another site in Texas has been drilled to 25,340 ft. If a fifth hole of 22,599 ft was drilled, where would it be placed in a list of these sites ordered from the least to the greatest depth?

1. Nelani is not sure what question the problem is asking her to solve.

   a. Tools   b. Solutions
   c. Checks   d. Questions

2. Ria has solved the word problem, but she is not sure her answer is correct.

   a. Tools   b. Solutions
   c. Checks   d. Questions

3. Les has started the word problem, but he is having difficulty placing the fifth number in the list.

   a. Tools   b. Solutions
   c. Checks   d. Questions

4. Gina understands which kind of answer the problem requires, but she is confused about the way to find it.

   a. Tools   b. Solutions
   c. Checks   d. Questions

Solve. Use the Help File if you need assistance.

5. The shortest possible surface distance from Moscow to Mexico City is 6,688 miles. From Mexico City to Rome is 6,353 miles, and from Rome to Tokyo, the surface distance is 6,124 miles. The distance from Tokyo to New York is 6,735 miles. The distance from New York to Beijing is 6,823 miles. What are the lengths of the two shorter routes to travel?

_____

Use with pages 8–9.

PRACTICE Addition and Subtraction

Add or subtract.

1. 5   2. 7   3. 8   4. 4   5. 9   6. 4
   −2   −3   −8   +1   −6   +7
   ───   ───   ───   ───   ───   ───
    3   4   0   5   3   11

7. 6   8. 3   9. 5   10. 2   11. 4   12. 7
   −0   +5   +9   −2   −3   +7
   ───   ───   ───   ───   ───   ───
    6   8   14   0   1   14

Write the missing number.

13. $3 + 5 =$ __8__   14. $6 + 8 =$ __14__   15. $4 + 8 =$ __12__   16. $5 + 9 =$ __14__

17. $7 + 4 =$ __11__   18. $8 + 6 =$ __14__   19. $8 - 4 =$ __4__   20. $9 - 6 =$ __3__

21. $7 +$ __8__ $= 15$ 22. ___ $+ 6 = 7$ 23. $3 -$ ___ $= 0$

24. $(4 +$ __0__ $) - 3 = 2$ 25. $3 +$ ___ $= 10$ 26. $6 + ($ ___ $+ 3) = 13$

27. ___ $- 7 = 9$ 28. ___ $+ 6 = 15$ 29. ___ $+ 5 = 5$

30. $(5 + 6) + 2 = 5 + (6 + 2) =$ ___   31. $8 + (3 +$ ___ $) = 3 + (9 + 8)$

32. $7 + (0 + 6) = 0 + ($ ___ $+ 6)$ 33. $15 + 14 = 14 +$ ___

Solve.

34. Haru and his friends are doing a survey on the number of years their parents have lived in the United States. Haru's parents immigrated from Japan 7 years ago. Maria's parents, both 36 years old, were born in the United States. Saul's parents came to the United States from Poland 14 years ago. How many years have their parents been in the United States altogether?

___

Use with pages 10–11.

PRACTICE — Front-End Estimation

Estimate the sum or the difference.

1.   3,902
       4,304
    + 6,100

2.   4,113
       5,499
    + 2,867

3. $14.70
       15.85
    + 19.99

4.   171,745
      402,923
    + 36,820

5.   4,615
    − 2,195

6.   14,624
    − 4,798

7.   65,593
    − 23,000

8. $670.11
    − 406.21

9.   62,707
     32,832
      4,050
    + 6,567

10. $325.17
      247.19
    +   45.60

11. 10,130
     4,123
     3,379
    + 9,807

12. 67,911
    800,526
    + 641,925

13. $126.45
    − 125.37

14. 40,060
    − 29,142

15. 7,671
    − 1,797

16. 7,116
    − 5,044

17. 46,000
     59,100
     17,600
    + 24,399

18. 17,183
    33,999
     4,876
    + 3,000

19. $ 5.73
      1.27
      3.14
    + 17.82

20. 4,905
    5,617
    + 3,812

21. $472.71
      39.65
    + 14.10

22. $424.95
    − 315.86

23. 167,009
    492,170
    + 248,698

24. 11,984
    − 10,960

25. 16,731
    − 6,426

26. $499.99
     562.98
    + 31.81

27. 65,631
    13,333
    + 4,100

28. 145,776
    392,902
    + 72,605

PRACTICE | Rounding and Estimating

Round to the nearest thousand.

1. 1,766 _____
2. 8,201 _____
3. 114,862 _____
4. 7,355 _____
5. 72,571 _____
6. 26,399 _____
7. 995,924 _____
8. 6,004 _____

Round to the nearest hundred thousand.

9. 362,209 _____
10. 221,696 _____
11. 755,261 _____
12. 479,800 _____
13. 843,356 _____
14. 925,555 _____

Round to the nearest dollar.

15. $1.96 _____
16. $18.55 _____
17. $7.22 _____
18. $34.08 _____
19. $0.74 _____
20. $26.32 _____
21. $0.41 _____
22. $51.77 _____

Estimate the sum or the difference.

23. 420 + 390
24. 65,782 + 32,021
25. $50.85 − $14.69
26. 545 + 691

27. 1,101 − 173
28. 83,590 − 18,900
29. 237 + 593
30. 11,500 + 6,700

31. $256.11 − $ 17.10
32. 420 − 118
33. $35.98 + $10.52
34. 486 + 709

35. 44,275 − 31,995
36. 9,794 + 215
37. 6,289 + 2,450
38. 353 + 217

Use with pages 14–15.

PRACTICE: Using Information from an Infobank

DISTANCES LIGHT TRAVELS (IN MILES)	
per second	186,000
per minute	11,160,000
per hour	669,600,000
per day	16,070,400,000

Write the letter of the correct answer.

1. What is the value of the digit *1* in the number of miles traveled by light per second?
   a. 1,000
   b. 10,000
   c. 101,000
   d. 100,000

2. What is the expanded form of the number 1,000,000,005?
   a. 1,000,000 + 500
   b. 1,000,000,00 + 5
   c. 1,000,000,005 + 105,000
   d. 1,000,000,000 + 5

Solve. Use the Infobank or an outside source if you need to.

3. How much farther does light travel in 1 hour than in 1 second?

4. Write the word name for the number of miles light travels in 1 day.

5. What is the mean distance in miles between the earth and the moon?

6. Does it take more than or less than a minute for light to travel from the moon to the earth?

7. How far does light travel in 1 day and 1 hour?

8. How much farther does light travel in 1 day and 1 hour than in 1 hour and 1 second?

9. If it takes the light from a certain star 1 day and 1 hour and 1 minute and 2 seconds to reach the eye of a lizard on a certain planet, how many miles from the star is the planet?

10. What is the mean distance in miles between the sun and the earth? Does it take light from the sun more or less than a minute to reach the earth?

Use with pages 16–17.

PRACTICE Adding Whole Numbers

Add.

1. 471,986 + 78,918
2. 197 + 26
3. 213,675 + 2,544
4. $325.67 + 236.32

5. 4,876 + 123
6. 433,785 + 267,430
7. $703.28 + 96.45
8. 270,487 + 539,734

9. 14,660 + 21,231
10. 97,478 + 2,077
11. 353,252 + 159,298
12. 78,453 + 14,879

13. 371,260 + 524,637
14. 478 + 398
15. 345,678 + 59,455
16. 3,216 + 4,182

17. 416 + 312
18. $462.37 + 124.82
19. $843.02 + 68.95
20. $267.53 + 429.64

21. 473 + 698 = _____

22. 7,894 + 4,786 = _____

23. 1,478 + 392 = _____

24. $475.55 + $3.55 = _____

25. 514,540 + 455,401 = _____

26. 419,737 + 398,495 = _____

Solve.

27. Sandra dreams each day of participating in the All-Sports Marathon. She has been saving her allowance for three months in order to buy sports equipment. Her monthly allowance is $20.00. List the four items she can buy with her savings.

   Hockey stick, $10.00   Baseball bat, $12.00

   Basketball, $12.00   Tennis racket, $90.00

   Baseball mitt, $10.00   Ice skates, $110.00

   Skateboard, $92.00   Warm-up jacket, $150.00

Use with pages 18–19.

PRACTICE — Column Addition

Add.

1. 3,184 + 1,213 + 679 = _____
2. 4,141 + 2,525 + 13,104 = _____
3. 2,341 + 3,838 + 1,066 = _____
4. 21,116 + 1,649 + 771 = _____
5. 6,010 + 264 + 917 = _____
6. 9,804 + 11,503 + 225 = _____

Add. Check your answers by adding up.

7. 205
    234
  + 221

8. $1.74
    2.66
  + 5.55

9. 6,035
    2,013
  + 5,600

10. 7,198
    2,007
  + 1,901

11. 124
    7,430
    1,098
  +   28

12. $ 8.42
    12.84
    51.27
  + 8.23

13. 1,098
    931
    85
  + 6,131

14. 17,001
    21,960
    840
  + 6,908

15. $769.43
    109.78
    56.80
  +   0.64

16.   13
    8,896
    371
  + 1,393

17. $246.89
    5.40
    27.62
  + 869.28

18. 325,936
    9,683
    14,872
  +   567

Solve.

19. Gail is paying her monthly bills. She pays $1,254.76 for her mortgage, $57.54 for her electric bill, $23.34 for her telephone bill, and $215.98 for her credit cards. How much has Gail paid so far? _____

20. Gail rides her bicycle each day after work. She rides 5 miles on Monday, 12 miles on Tuesday, 18 miles on Wednesday, 9 miles on Thursday, and 22 miles on Friday. How many miles has Gail ridden this week? _____

PRACTICE: Subtracting Whole Numbers

Subtract.

1. 17 − 13
2. 27 − 12
3. 632 − 511
4. 97 − 78

5. 3,782 − 461
6. 4,832 − 2,763
7. $67.37 − 49.53
8. 1,984 − 96

9. 498 − 375
10. $97.83 − 28.96
11. 3,794 − 1,886
12. $76.82 − 34.69

13. 817 − 433
14. 3,967 − 489
15. 425 − 275
16. 43 − 29

17. 755 − 472 = _____
18. 5,984 − 4,726 = _____

19. $57.25 − $23.56 = _____
20. 695,450 − 475,350 = _____

21. 34,790 − 3,740
22. 96,378 − 496
23. 96,378 − 42
24. $1,776.54 − 254.41

25. $823.42 − 459.37
26. 43,376 − 21,482
27. 35,632 − 14,751
28. $752 − 633

29. 979,631 − 48,732
30. 793,456 − 724,538
31. $4,718 − 3,609
32. 349,136 − 549

33. 621,891 − 437,674
34. $3,648.81 − 439.10
35. $7,513 − 625
36. 991,734 − 832,476

Use with pages 22–23.

PRACTICE: Subtracting With Zeros

Subtract. Check your answers by adding.

1. 800 − 645 = _____
2. 9,002 − 386 = _____
3. 720 − 435 = _____
4. 5,400 − 3,257 = _____
5. 375,000 − 296,020 = _____
6. 55,000 − 2,890 = _____

7. 302 − 281
8. 400 − 365
9. 703 − 275
10. 570 − 298

11. 9,070 − 8,350
12. 39,002 − 697
13. 41,000 − 398
14. 7,050 − 329

15. 3,004 − 731
16. 7,005 − 2,186
17. 70,320 − 8,579
18. 37,000 − 9,755

19. 400 − 216
20. 870,000 − 87,937
21. 604,200 − 355
22. 20,000 − 19,725

23. 300 − 258
24. 39,100 − 678
25. 4,700 − 3,505
26. 6,701 − 582

27. 9,000 − 725
28. 6,700 − 5,988
29. 70,400 − 1,798
30. 40,600 − 1,728

Solve.

31. When the results of the school election were posted, some of the numbers were missing. Copy and complete the table to show how many votes each candidate had. Who won the election?

Candidate	Grade 6	Grade 7	Grade 8
Sue	93		208
Clifford		109	
Total	304	217	270

Use with pages 24–25.

PRACTICE — Estimation

Shoes
loafers $17.95
sneakers $14.95

Socks
anklets $1.39
knee socks $2.19

Blouses
long sleeve $8.45
sleeveless $6.50

Slacks (colors)
solid $21.50
plaid $23.75

Use the information from the list for each problem. Decide whether you should find an *estimate* or *exact* answer.

1. Ann wants to buy a pair of loafers and a pair of knee socks. She has $25. Does she have enough money? _____

2. Natalie bought a sleeveless blouse and a pair of blue slacks. She paid for them with two $20 bills and received $12.00 change. Was that the right amount of change? _____

3. Sandy bought a pair of loafers and a pair of plaid slacks. Amy bought a pair of sneakers and a pair of green slacks. About how much more or how much less did Sandy's outfit cost? _____

Solve. Estimate your answer. Find the exact answer *only* if you need to.

	Suppose Myrna had	Can she buy	
4.	$18.50	a pair of loafers and a pair of anklets	_____
5.	$5.00	a pair of anklets and a pair of knee socks	_____
6.	$20.10	a pair of loafers and a pair of knee socks	_____
7.	$36.75	a sleeveless blouse, a pair of plaid slacks, and 2 pair of knee socks	_____
8.	$50.00	the least expensive of each item	_____

Use with pages 26–27.

PRACTICE Equations

Write the missing number.

1. $10 + n = 16$
   n = __6__

2. $12 + n = 24$
   n = __12__

3. $14 - n = 8$
   n = _____

4. $11 - n = 5$
   n = __6__

5. $n + 4 = 21$
   n = __17__

6. $n - 1 = 17$
   n = _____

7. $15 - n = 10$
   n = __5__

8. $12 + n = 19$
   n = __7__

9. $n + 7 = 9$
   n = _____

10. $5 - n = 3$
    n = __2__

11. $n - 6 = 12$
    n = __18__

12. $4 + n = 9$
    n = _____

13. $n + 10 = 13$
    n = __3__

14. $4 - n = 0$
    n = __4__

15. $7 + n = 19$
    n = _____

16. $2 + n = 15$
    n = __13__

17. $n - 8 = 5$
    n = __13__

18. $n - 9 = 9$
    n = _____

19. $21 - n = 21$
    n = __1__

20. $8 + n = 9$
    n = __1__

21. $n - 7 = 1$
    n = _____

Solve.

22. Candy and Denise played two games of table tennis. When 3 is subtracted from the number of points Candy scored in the first game, the difference is 12. How many points did Candy score?
    __15__

23. The sum of the number of points that Denise scored in two games is 15. How many points did Denise score in the second game if she scored 5 points in the first game?
    _____

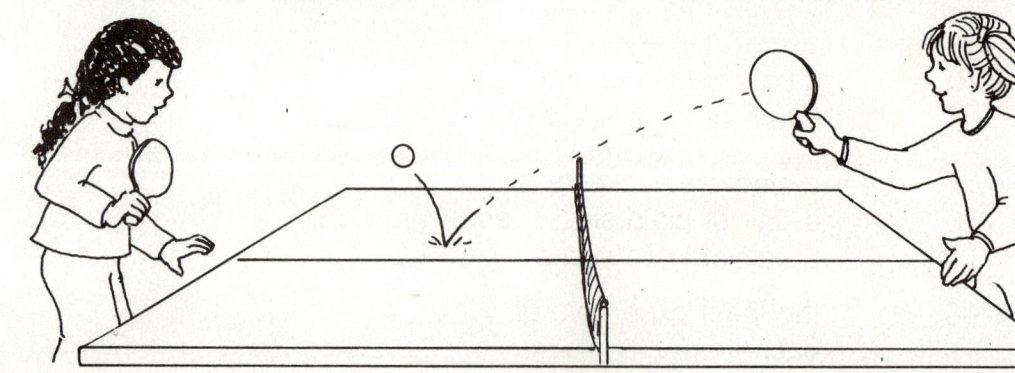

PRACTICE — Tenths and Hundredths

Write as a decimal.

1. 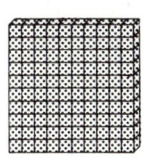 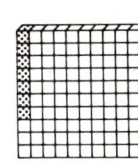   .1 10
   .01 Hun
   .001

2.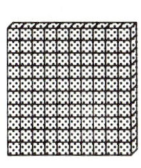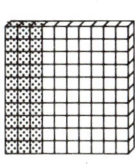

3. $6\frac{7}{10}$ = 6.07

4. $8\frac{1}{100}$ = 8.0001

5. $4\frac{1}{100}$ = 4.001

6. $3\frac{3}{100}$ = 3.003 3.03

7. $5\frac{1}{10}$ = 5.01

8. $\frac{7}{10}$ = .07

9. $2\frac{1}{100}$ = 2.001 2.01

10. $41\frac{1}{10}$ = 41.01

11. $\frac{7}{100}$ = .007

12. $6\frac{7}{100}$ = 6.007 6.07

13. $4\frac{8}{10}$ = 4..8

14. $11\frac{99}{100}$ = 11.0099

15. seven tenths   7/10

16. one and one tenth   1 1/10

17. five and two tenths   5 2/10

18. six and three tenths   6 3/10

19. eight and three hundredths   8 3/100

20. seven and twenty hundredths   7 20/100

21. nine and two hundredths   9 2/100

22. four and six hundredths   4 6/100

23. seventeen and thirty-four hundredths   17 34/100

24. nine and two tenths   9 2/10

Write the word name.

25. 0.7   7/10 seven tenths

26. 0.04   4/100 4 hundreths

27. 3.17   3 17/100

28. 6.02 _____

Use with pages 38–39.

PRACTICE Hundred-Thousandths

Write in expanded form.

1. 3.65762 _____
2. 6.273 _____
3. 49.061 _____
4. 812.793 _____

Write each as a decimal.

5. four hundred one thousandths _____
6. four thousandths _____
7. ninety-two and four tenths _____
8. eighty-seven thousandths _____
9. $6\frac{202}{1,000}$ _____
10. $7\frac{207}{1,000}$ _____
11. $37\frac{9}{100}$ _____

Copy the place-value chart below. Write each number as a decimal in the chart.

12. forty-one and six tenths
13. nine hundred and two hundred sixty-one thousandths
14. seventy-six and seven ten-thousandths
15. $656\frac{1}{1,000}$
16. one and sixty-two hundred-thousandths
17. $24\frac{7}{1,000}$
18. $31\frac{412}{1,000}$

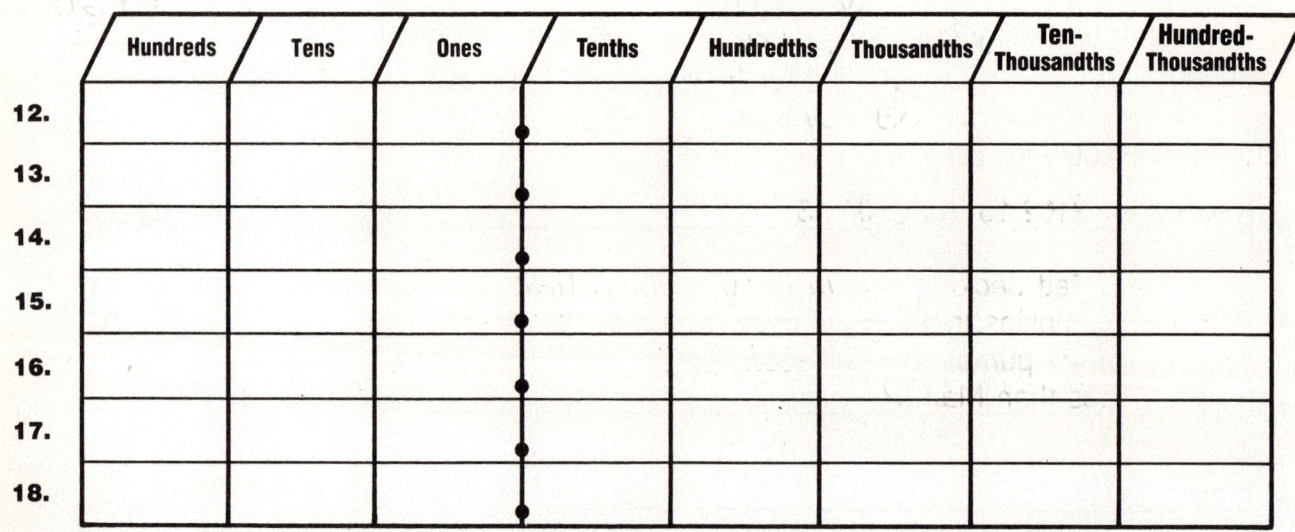

	Hundreds	Tens	Ones	Tenths	Hundredths	Thousandths	Ten-Thousandths	Hundred-Thousandths
12.								
13.								
14.								
15.								
16.								
17.								
18.								

PRACTICE: Checking for a Reasonable Answer

Read each problem. Without finding the exact answer, write the letter of the correct answer.

1. If 2 pumpkins together contain 678 seeds, and the smaller contains only 235 seeds, approximately how many seeds does the larger contain?

   a. 550   b. 450   c. 300

2. Matt and Jason are selling pumpkins at their town fair. They start the day with 423 pumpkins, and end the day with 97. How many did they sell during the fair?

   a. 326   b. 257   c. 357

3. Jason and Matt sell medium-size pumpkins for $0.37 per pound. How much would they charge for 5 medium-size pumpkins, each of which weighed 1.25 pounds?

   a. $1.85   b. $2.31   c. $23.10

4. If 3 pounds of roasted pumpkin seeds cost $3.75, what is the approximate cost of 9 pounds?

   a. $11.00   b. $27.00   c. $33.00

5. A man brings 3 large and 2 small pumpkins to the cash register. The large pumpkins, each weighing 5.5 pounds, sell for $0.31 per pound. The small pumpkins, each weighing 0.75 pounds, sell for $0.55 per pound. How much does the man pay?

   a. $2.03   b. $4.65   c. $5.95

6. Matt and Jason's largest pumpkin weighed 102 pounds less than the pumpkin that won the contest for the largest squash. Their pumpkin weighed 309 pounds. How much did the winning pumpkin weigh?

   a. 457 lb   b. 411 lb   c. 500 lb

7. Jason and Matt sell $43.13 worth of pumpkins and roasted seeds between 9:00 A.M. and 10:00 A.M., and $65.00 worth between 10:00 A.M. AND 11:00 A.M. Rent for the booth is $6.00 for 2 hours. How much profit did they earn between 9:00 A.M. and 11:00 A.M.?

   a. $84.13   b. $102.13   c. $108.13

8. Jason and Matt have to drive about 82 miles to get home. On the way, they stop for dinner. How long does it take them to get home if they average 37 miles per hour and only spend 45 minutes for dinner?

   a. $1\frac{1}{2}$ hours   b. 2 hours   c. 3 hours

9. Jason and Matt decide to carve some of the 97 pumpkins that are left over. Matt carves 21 pumpkins and Jason carves 3 less than Matt. How many pumpkins are still uncarved?

   a. 58   b. 65   c. 71

10. Matt buys 3 jars of apple butter and 2 jars of raspberry jam at the fair. Apple butter costs $2.79 per jar and jam costs $1.88 per jar. How much money did Matt spend?

    a. $10.52   b. $12.13   c. $20.65

Use with pages 42–43.

PRACTICE: Comparing and Ordering Decimals

Write >, <, or = for each ◯.

1. 0.6987 ◯ 1.76
2. 88.9 ◯ 88.90
3. 11.15 ◯ 11.1500
4. 4.768 ◯ 4.867
5. 67.90765 ◯ 67.97065
6. 20.99 ◯ 20.9
7. 100.001 ◯ 101.000
8. 2.9760 ◯ 2.760
9. 34.2525 ◯ 34.2552

Write in order from the least to the greatest.

10. 5.1, 5.103, 5.001, 5.0001 _____
11. 1.07, 1.0719, 0.971, 1.00156 _____
12. 0.5589, 1.5589, 0.95589, 0.95 _____

Write in order from the greatest to the least.

13. 7.635, 6.735, 6.753, 6.635 _____
14. 2.7, 2.765, 0.9932, 0.2 _____
15. 0.31477, 0.3131, 1.3131, 0.404 _____

Write a number for each point on the line that is marked by a letter.

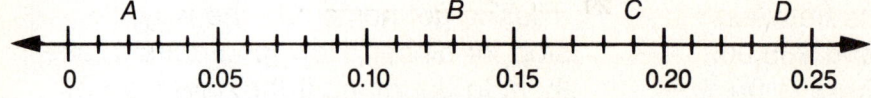

16. A _____   17. B _____   18. C _____   19. D _____

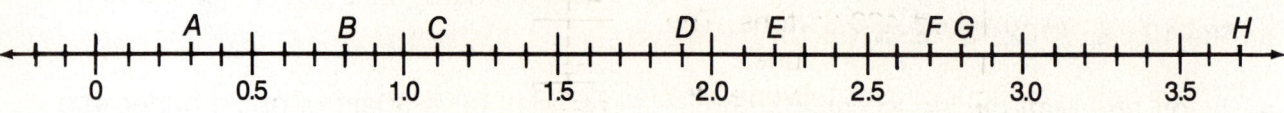

20. A _____   21. B _____   22. C _____   23. D _____
24. E _____   25. F _____   26. G _____   27. H _____

Use with pages 44–45.

PRACTICE Rounding Decimals

Round to the nearest whole number.

1. 4.09 _____
2. 69.9 _____
3. 33.33 _____
4. 27.7 _____
5. 23.01 _____
6. 4.87 _____
7. 0.932 _____
8. 14.61 _____
9. 17.449 _____

Round to the nearest tenth.

10. 16.04 _____
11. 435.27 _____
12. 86.319 _____
13. 1.99 _____
14. 76.66 _____
15. 39.8544 _____
16. 111.655 _____
17. 9.06 _____
18. 4.6666 _____

Round to the nearest hundredth.

19. 73.479 _____
20. 8.99199 _____
21. 0.5051 _____
22. 14.4444 _____
23. 80.1666 _____
24. 8.688 _____

Round to the nearest thousandth.

25. 30.6991 _____
26. 80.67841 _____
27. 0.00701 _____

Round to the nearest ten-thousandth.

28. 82.17959 _____
29. 67.08989 _____

Solve.

30. Garfield's Canning Company ordered 860 tons of tin from a mining company. The order came in two separate loads. The first load weighed 420.65 tons. The second load weighed 439.45 tons. Estimate the weight of the deliveries to the nearest whole number.

    First load _____

    Second load _____

    Did Garfield's get all the tin they ordered? _____

Use with pages 46–47.

PRACTICE: Making a List

Madrigal Jones is having a story-book costume party. Everybody who comes must dress as a character from a story. Make an organized list and solve each problem.

1. Before the party, The Mad Hatters and the March Hares decide to play a game of kickball. If the entire number of runs scored in the game is 8, how many different possible scores could this game have had? _____

2. The Mad Hatters decide to form a decorating committee. Of the 11 Mad Hatters at the party, 6 are boys. How many different groups containing 5 male Mad Hatters can be formed? _____

3. There are 3 Dormice, 3 White Rabbits, and 3 Caterpillars at the party. They are teaming up to play jump-rope. How many different teams can be formed containing one Dormouse, one White Rabbit, and one Caterpillar? _____

4. The 4 Queen of Hearts are also in the sack race. If none of the Queens tie, in how many different orders can they cross the finish line? _____

Use with pages 48–49.

PRACTICE: Making a List

Make an organized list and solve.

5. Another group of Mad Hatters forms the refreshment committee. How many different groups that each contain one boy and one girl can be formed from among the 6 boy Mad Hatters and the 5 girl Mad Hatters? _____

6. There are 3 Knaves of Heart at the party. If each of the 4 Queens of Hearts challenges each of the knaves to a game of ticktacktoe once during the party, and each of the knaves challenges each Queen to a game of ticktacktoe once, how many games occur between the knaves and the queens throughout the party? _____

7. There are 4 Alices at the party. Each Alice plays a game of checkers against every other Alice. How many games of checkers are played between the Alices at the party? _____

8. Each of the 4 Alices thinks of a number from 1 to 10. To everyone's surprise, it turns out they are all thinking of the same number! In how many different ways can this happen? _____

9. The 3 Dormice take part in a game of tag. One of the 2 team captains is Amy Chung; the other is Tony Madden. In how many different ways can the Dormice be split up between Amy's and Tony's teams? _____

Use with pages 48–49.

PRACTICE: Estimating Decimal Sums and Differences

Estimate. Write > or < for each ◯.

1. 8.45 + 4.89 ◯ 12
2. 15.29 + 18.624 ◯ 34
3. 35.917 + 4.3 ◯ 40
4. 3.32 + 1.8 + 5.711 ◯ 10
5. 0.75 + 1.3 + 0.827 ◯ 3
6. 15.2 + 2.5 + 2.17 ◯ 20
7. 2.8 + 0.895 + 0.92 ◯ 4
8. 6 − 5.23 ◯ 1
9. 10.16 − 5.4 ◯ 5
10. 29.75 − 20.36 ◯ 10
11. 49.427 − 19.8 ◯ 30
12. 100 − 98.6 ◯ 2
13. 83.625 − 70.5 ◯ 13
14. 8.32 − 0.905 ◯ 7

Estimate.

15. 2.65 + 5.872

16. 18.31 + 10.95

17. 5.14 + 0.95 + 0.895 + 3.01

18. 2.81 + 3.017 + 4.59

19. 8.46 − 2.95

20. 18.0 − 7.56

21. 60.88 − 39.39

22. 12.73 − 9.8

Solve.

23. Below are the results of the 1976 Presidential election. About how many total votes were cast?

_____

24. In 1850, the average life expectancy for females was 40.5 years. By 1950, it had risen to 72.03 years. About how much longer could a female born in 1950 expect to live than one born in 1850?

_____

Candidate	Votes
Carter (Dem.)	40.8 million
Ford (Rep.)	39.1 million
McCarthy (Ind.)	0.7 million
Total	?

Use with pages 50–51.

PRACTICE: Adding with Decimals

Find the sum.

1. 67.381
   + 53.400

2. 1.6998
   + 8.5

3. 43.1
   + 7.855

4. 20.77
   + 59.67

5. 32.65
   + 29.71

6. 20.766
   + 35.98

7. 4.17
   + 11.86

8. 25.64
   + 33.8

9. 30.981
    25.64
   + 44.778

10. 70.38
    15.64
   + 23.75

11. 67.09
    14.85
   + 58.77

12. 9.560
    2.374
   + 5.883

13. 11.05 + 5.44 = _____

14. 32.46 + 24.956 = _____

15. 4.54 + 3.55 = _____

16. 81.091 + 14.75 + 5 = _____

17. 85.98 + 76.6 = _____

18. 85.81 + 75.76 + 13 = _____

19. 40.16 + 3.46 = _____

20. 65.383 + 25.618 = _____

21. 6.381 + 3.763 = _____

22. 78.817 + 57.06 + 22 = _____

Hector, Sally, and Jennifer went out to lunch to celebrate Hector's birthday. The table shows how much each part of their lunch cost.

23. Copy and complete the table.

24. What was the total cost of all three lunches?

   _____

25. If Jennifer paid for Hector's lunch as well as her own, how much did she pay?

   _____

	Hector	Sally	Jennifer
Salad	$1.50	$1.25	$0.75
Sandwich	$2.10	$2.99	$1.90
Drink	$0.95	$0.95	$0.65
Total			

Use with pages 52–53.

PRACTICE: Subtracting with Decimals

Subtract.

1. 64.9 − 38.1	2. 7.3 − 5.66	3. 6.4 − 2.561	4. $73.04 − 71.08
5. $10.14 − 7.22	6. 77.60 − 32.98	7. 3.33 − 2.34	8. 21.74 − 1.8
9. 76.835 − 42.0079	10. 9.6 − 8.5973	11. 12.3704 − 6.8	12. 13.865 − 12.977
13. 37.0031 − 28.9472	14. 4.73 − 2.2681	15. 57.8378 − 7.84	16. 2.3524 − 1.6107

17. 8.46 − 3.58 = _____

18. $7.69 − $4.90 = _____

19. 7.67 − 5.3777 = _____

20. 13.46 − 7.1903 = _____

21. 3.4771 − 2.8385 = _____

22. 9.012 − 8.7654 = _____

23. 14.38 − 6.75 = _____

24. 50.05 − 48.721 = _____

25. $87.15 − $70.28 = _____

26. 3.6478 − 1.3156 = _____

27. 3.81 − 2.9702 = _____

28. $40.56 − $17.48 = _____

29. 12.713 − 10.855 = _____

30. $3.90 − $1.97 = _____

Solve.

31. Monrovia, Liberia, receives more rain than any other city in the world. It receives 202.01 inches annually. Antofagasta, Chile, receives the least rain of any city in the world. It gets 0.02 inches annually. How many more inches of rain fall on Monrovia than on Antofagasta?

_____

Use with pages 54–55.

PRACTICE: Choosing the Operation

	Game 1	Game 2	Game 3	Game 4	Game 5
Points scored by Jack	7	13	11	6	17

Use the information above to solve. Write the letter of the correct operation.

1. What was the difference between Jack's point total in his highest and lowest scoring games?

   a. addition   b. subtraction   c. both addition and subtraction

2. How many more points did Jack score in his third, fourth, and fifth games than in his first and second games?

   a. addition   b. subtraction   c. both addition and subtraction

Solve.

	Game 1	Game 2	Game 3
Points scored by Flyers	40	35	37
Points scored by Pistons	33		
Points scored by Warriors		31	
Points scored by Stars			49

In the third game, the Flyers led 20 to 17 at halftime.

3. How many points did the Flyers score in their first 2 games of the season? _____

4. How many more points did the Flyers score in their first 2 games than their opponents? _____

5. How many more points were scored by the Stars than by the Flyers in the second half of their game? _____

6. In the third game, how many points were scored in the first half of the game? _____

Use with pages 56–57.

PRACTICE — Identifying Extra/Needed Information

Bobby and Ray are on the Arness School basketball team. On Saturday, they played the team from Birch School.

Write the letter of the answer that best describes each problem.

1. The Arness team won with a score of 64 points. Bobby scored 12 points and Ray scored 8 points. How many points in total were scored by both teams?

   a. There is not enough information to solve the problem.

   b. There is more than enough information to solve the problem.

2. The Arness team's high scorer had 15 points, and the Birch team's high scorer had 13 points. The Birch team lost by 4 points. Which team had the highest single scorer?

   a. There is not enough information to solve the problem.

   b. There is more than enough information to solve the problem.

Solve if you have enough information. If not, write *not enough information*.

3. At the end of the first half, Birch was leading by 4 points. Arness had 30 points. How many points did Birch have? _____

4. The attendance at the game was 217 people. If the adult admission charge was $1.50 and the student admission charge was 75¢, how much money was taken in at the box office? _____

5. The Arness team's record for the previous season was 14 wins and 9 losses. They played Birch 5 times. How many times did they win against Birch? _____

6. Arness scored 64 points during their first game of the season; 67 during the second; and 58 points during the third. During their first 3 games, Birch scored 187 points. Which team leads in points after the first 3 games? _____

7. Ray scored 8 points during his first game; 4 points more than that during his second game; and one point less than his first game during his third game. How many points did he score during his first 3 games? _____

8. During last year's season, attendance for all Arness home games was 1,250. After half of this year's season, it has reached 670. Attendance for Birch home games last season was 1,190. Which team has a greater attendance record so far this year? _____

PRACTICE: Multiplication Facts and Properties

Multiply.

1. $6 \times 7 =$ _____
2. $5 \times 9 =$ _____
3. $(3 \times 4) \times 5 =$ _____
4. $7 \times (4 \times 3) =$ _____
5. $4 \times (5 \times 3) =$ _____
6. $2 \times 0 =$ _____

7. 7×9
8. 2×9
9. 5×9
10. 7×1
11. 8×5

12. 3×6
13. 7×4
14. 3×8
15. 8×0
16. 8×6

17. $9 \times 4 =$ _____
18. $(9 \times 6) \times 0 =$ _____
19. $7 \times (1 \times 6) =$ _____
20. $(4 \times 1) \times 1 =$ _____
21. $6 \times 9 =$ _____
22. $8 \times 8 =$ _____

23. 6×6
24. 2×6
25. 6×7
26. 8×1
27. 4×4

28. 0×5
29. 1×3
30. 8×2
31. 5×5
32. 7×2

33. 6×4
34. 9×9
35. 1×0
36. 3×7
37. 3×3

Write the missing number. Then name the property used.

38. $1 \times 5 =$ _____

39. $3 \times 9 = 9 \times$ _____

40. $4 \times (7 + 1) = ($ _____ $\times 7) + (4 \times 1)$

41. $8 \times 3 =$ _____ $\times 8$

42. $7 \times 0 =$ _____

43. $3 \times (3 \times 6) = ($ _____ $\times 3) \times 6$

Use with pages 70–71.

PRACTICE: Multiplying with Multiples of 10; 100; 1,000

Multiply.

1. 70 × 10 = _____
2. 800 × 100 = _____
3. 1,000 × 500 = _____
4. 10 × 99 = _____
5. 100 × 525 = _____
6. 300 × 3,000 = _____
7. 200 × 10 = _____
8. 100 × 150 = _____
9. 1,000 × 110 = _____

10. 10 × 4
11. 900 × 90
12. 50 × 5
13. 200 × 70

14. 300 × 80
15. 90 × 8
16. 7,000 × 9,000
17. 30 × 5

18. 700 × 500
19. 400 × 500
20. 1,000 × 6,000
21. 1,000 × 8,000

22. 5,000 × 200
23. 600 × 100
24. 4,000 × 300
25. 8,000 × 5,000

Solve.

26. The radio announcer explains that the first listener to call the winning telephone number will win five free concert tickets. To find the winning telephone number, copy the letters and answer rules. Multiply. Then write the number of zeros in each product above the corresponding lines.

___ ___ ___   ___ ___ ___ ___
 A   B   C   D   E   F   G

A. 20,000 × 30 =
B. 1,000 × 800 =
C. 100,000 × 9 =
D. 40,000 × 200 =
E. 100 × 4 =
F. 3,000 × 300,000 =
G. 10,000 × 7,000 =

PRACTICE: Estimating Products

Estimate. Then write > or < to show how you would adjust the estimate.

1. 379 × 192 _____
2. 41 × 112 _____
3. 632 × 510 _____
4. 79 × 27 _____

5. 63 × 62 _____
6. 295 × 173 _____
7. 91 × 91 _____
8. 395 × 77 _____

9. 222 × 416 _____
10. 222 × 123 _____
11. 531 × 62 _____
12. 428 × 317 _____

13. 666 × 66 _____
14. 571 × 86 _____
15. 94 × 689 _____
16. 431 × 327 _____

Estimate in two different ways.

17. 12 × 45
_____
_____

18. 22 × 16
_____
_____

19. 78 × 26
_____
_____

20. 43 × 46
_____
_____

Solve.

Products, Incorporated, manufactures three models of its Product Producer, a machine that multiplies numbers. Model A can give products less than 1,000. Model B gives products greater than 1,000 but less than 5,000, and Model C gives products greater than 5,000. Which model of Product Producer could find

21. 23 × 127? _____
22. 37 × 18? _____
23. 271 × 383? _____

Use with pages 74–75.

PRACTICE Writing an Equation

Each year there is a frog-jumping contest at the Marion Town Fair. Each entry takes 3 jumps; only the longest one counts. The winner this year was Mr. Belly. Second place went to Superfrog, and third place was won by Frisco Fred.

Read each problem. Write the letter of the correct equation.

1. Mr. Belly's first 2 jumps were 62.4 and 87.5 centimeters long. The combined length of all 3 jumps was 267.2 centimeters. What was the length of Mr. Belly's best jump?

   a. $87.5 + 62.4 + 267.2 = n$

   b. $87.5 + 62.4 + n = 267.2$

   c. $87.5 + 62.4 - 267.2 = n$

2. Superfrog's longest jump was 103.4 centimeters. During practice, however, his best jump was 112.3 centimeters. How much longer was Superfrog's best jump ever than his best jump at the fair?

   a. $103.4 + d = 112.3$

   b. $112.3 \times d = 103.4$

   c. $d - 112.3 = 103.4$

Write an equation and solve.

3. Frisco Fred's best jump was his first jump of 101.9 centimeters. His other 2 jumps were 87.8 and 95.2 centimeters. What was the total distance covered by Frisco Fred in all 3 jumps?

4. Of the 43 frogs entered in the contest, 34 were bull frogs, 1 was a tree frog, and the rest were golden-leopard frogs. What was the number of golden-leopard frogs in the contest?

5. In addition to Superfrog's 103.4-centimeter jump, he also had a jump of 100.5 centimeters. The combined length of all three of his jumps was 257.8 centimeters. How long was Superfrog's third jump?

6. Mr. Belly won with a jump of 117.3 centimeters. Frisco Fred's third-place distance was 101.9 centimeters. Superfrog finished 1.5 centimeters ahead of Frisco Fred. How much farther was the winning distance than the second-place distance?

PRACTICE: Multiplying by 1-Digit Factors

Multiply.

1. 143 × 7
2. 726 × 3
3. 4,891 × 6
4. 200 × 2

5. 8,578 × 8
6. $26.33 × 4
7. $80.14 × 6
8. $6.38 × 2

9. $59.02 × 8
10. $15.30 × 5
11. 697 × 8
12. 2,076 × 3

13. $41.87 × 6
14. 302 × 9
15. 42.30 × 6
16. $5.23 × 4

17. 9,024 × 5
18. $78.91 × 7
19. 37.66 × 2
20. $777.77 × 9

21. 4 × 7,931 = _____
22. 8,178 × 2 = _____
23. 9 × $98.01 = _____
24. 9 × 6,085 = _____
25. 5 × $6.22 = _____
26. 4 × $10.08 = _____
27. 4 × 2 × 3,584 = _____
28. (3 × 3) × 8,984 = _____
29. 3 × (4,213 × 5) = _____
30. 4 × (2,516 × 2) = _____
31. 3 × (7 × 95) = _____
32. 5 × (6 × 1,004) = _____

33. $16.60 × 5
34. 232,755 × 4
35. 4,027 × 7
36. $368.15 × 2
37. 9,005 × 3

Use with pages 78–79.

PRACTICE: Multiplying by 2-Digit Factors

Multiply.

1. 41 × 83 = _____
2. 7,500 × 59 = _____
3. 45 × $9.20 = _____
4. 185 × 37 = _____
5. 41,622 × 59 = _____
6. 22 × 22 = _____
7. 918 × 77 = _____
8. 3,821 × 10 = _____
9. 243 × 23 = _____

10. 325 × 73
11. 67 × 25
12. $25.81 × 38
13. 428 × 41
14. $1.98 × 11

15. 45 × 13
16. $42.35 × 16
17. 1,275 × 15
18. $0.92 × 24
19. 1,584 × 12

20. $6.27 × 11
21. $42.10 × 22
22. 2,912 × 33
23. 198 × 13
24. $4.85 × 25

25. $0.25 × 32
26. $35.05 × 19
27. 4,814 × 27
28. 551 × 88
29. 151 × 43

Solve.

30. Quincy works in a record store. One day he sold 98 copies of a popular album in one hour. Each album costs $8.97. How much money was spent on that album in one hour? _____

31. Quincy unpacks a new shipment of records. The records are packed 36 to a box. There are 24 boxes in the shipment. How many records does Quincy have to unpack? _____

PRACTICE: Multiplying by 3-Digit Factors

Multiply.

1. 728 × 339
2. 364 × 716
3. 917 × 422
4. 204 × 638

5. 271 × 583
6. 953 × 426
7. 406 × 789
8. 378 × 115

9. 6,697 × 312
10. 926 × 426
11. 4,187 × 516
12. 2,350 × 709

13. $5.03 × 962
14. 9,325 × 182
15. $69.46 × 278
16. 159 × 159

17. 1,000 × 496
18. 2,564 × 321
19. $84.53 × 917
20. $5.70 × 336

21. 39.45 × 394
22. 4,224 × 666
23. $18.27 × 431
24. $2.35 × 400

25. 824 × 391 = _____
26. 540 × 1,628 = _____
27. 651 × 472 = _____
28. 3,892 × 715 = _____
29. 953 × 577 = _____
30. 2,001 × 333 = _____

Use with pages 82–83.

PRACTICE: Exponents and Squares

Write in exponent form.

1. two squared _____
2. seven to the fourth power _____
3. five to the sixth power _____
4. nine to the eleventh power _____
5. three to the fifth power _____
6. four to the eighth power _____
7. $2 \times 2 \times 2$ _____
8. $5 \times 5 \times 5 \times 5 \times 5 \times 5 \times 5$ _____
9. $8 \times 8 \times 8 \times 8$ _____
10. 10×10 _____
11. $6 \times 6 \times 6 \times 6 \times 6$ _____
12. $4 \times 4 \times 4$ _____
13. $3 \times 3 \times 3 \times 3 \times 3$ _____
14. $7 \times 7 \times 7 \times 7$ _____
15. $10 \times 10 \times 10$ _____
16. $2 \times 2 \times 2 \times 2 \times 2 \times 2 \times 2$ _____
17. $6 \times 6 \times 6 \times 6 \times 6 \times 6$ _____
18. 10 _____
19. $4 \times 4 \times 4$ _____
20. five to the fifth power _____

Write the product.

21. $6^2 =$ _____
22. $10^4 =$ _____
23. $8^1 =$ _____
24. $7^3 =$ _____
25. $3^4 =$ _____
26. $9^2 =$ _____
27. $5^2 =$ _____
28. $12^2 =$ _____
29. $8^2 =$ _____
30. $7^1 =$ _____
31. $4^3 =$ _____
32. $10^5 =$ _____
33. $5^3 =$ _____
34. $11^2 =$ _____
35. $1^{10} =$ _____
36. $6^4 =$ _____
37. $2^7 =$ _____
38. $3^5 =$ _____
39. $7^2 =$ _____
40. $5^6 =$ _____
41. $4^5 =$ _____

PRACTICE Estimation

Write the letter of the correct answer.

1. An orchard has 200 peach trees and 250 apple trees. The average peach tree produces about 22 bushels of peaches and the average apple tree produces about 19 bushels of apples. Which is the best estimate of the total number of bushels of fruit produced by the peach and apple trees?

   a. 8,000 bushels
   b. 9,000 bushels
   c. 10,000 bushels

2. There are 8 different varieties of fruit trees in the orchard. There are about the same number of trees for each variety. Mike believes there are 197 plum trees. About how many fruit trees are there in the orchard?

   a. 1,600
   b. 160
   c. 800

Estimate to solve. Explain your answer.

3. The orchard has 265 pear trees. About 3 gallons of plant food is used to feed each tree. Will 900 gallons of plant food be enough to feed all the pear trees?

4. One gallon of plant food costs $1.67. Will $1,200 be enough to buy 800 gallons of plant food?

5. A bushel of apples weighs about 37 pounds. A pickup truck can carry 1,200 pounds. Can the truck carry 25 bushels of apples at one time?

6. A bushel of apples costs $8.50 and a bushel of peaches costs $15.95. Will $40.00 be enough to buy 2 bushels of apples and a bushel of peaches?

7. It costs approximately $32 to care for each of the 1,152 trees in an orchard. Will $35,000 be enough to pay the expense of caring for the trees?

8. A tractor pulls a wagon that holds 50 bushels of pears. A bushel of pears weighs about 28 pounds. About how many pounds of pears can the tractor carry?

Use with pages 86–87.

PRACTICE: Multiplying Decimals by 10; 100; and 1,000

Multiply.

1. 0.5870 × 10
2. 0.04629 × 1,000
3. $0.25 × 100
4. 6.78 × 10
5. 0.0132 × 10

6. 0.3 × 1,000
7. $95.22 × 1,000
8. $86.37 × 10
9. 0.451 × 100
10. 57.1 × 100

11. 0.4 × 100
12. 649.2 × 1,000
13. $94.59 × 100
14. 75.01 × 1,000
15. $76.82 × 1,000

16. 100 × 71.29 = _____
17. 10 × 531.4 = _____
18. 1,000 × $4.06 = _____
19. 100 × 86.27 = _____
20. 10 × $239.68 = _____
21. 10 × $47.44 = _____
22. 100 × 67.444 = _____
23. 1,000 × 936.2 = _____
24. 10 × 32.58 = _____
25. 10 × 5.4007 = _____
26. 1,000 × 0.049 = _____
27. 100 × $3.89 = _____
28. 100 × $20.00 = _____
29. 10 × 0.0072 = _____
30. 1,000 × $0.05 = _____
31. 10 × 0.9401 = _____
32. 100 × 866.3 = _____
33. 1,000 × $7.22 = _____
34. 10 × $642.92 = _____
35. 10 × 0.003 = _____

Solve.

36. Bob is a TV programmer for a local cable station. He gives 3.7 minutes to each commercial segment shown by the station. If the station charges $1,000 per minute for showing commercials, how much does it earn for each segment?

_____

36   Use with pages 88–89.

PRACTICE: Solving Two-Step Problems

Complete the plan by writing the appropriate step.

1. The community center wants to raise $100,000 for a new roof. After $58,679 had been raised, a local company donated $3,500. How much money did the community center then need to reach its goal?

   Step 1: Find the total amount of money raised.

   Step 2: _____

2. George and his friends want to help raise money. They decide to build and sell bird feeders. George and 8 of his friends will each build 10 bird feeders. They plan to sell each bird feeder for $6.95. How much do George and his friends hope to raise?

   Step 1: _____

   Step 2: Find the total amount of money the boys hope to raise.

Solve.

3. Tonya and 2 of her friends start a child-care group. They charge $0.75 per child per hour. On Saturday morning, 8 children joined their group for 5 hours. How much money did they make? _____

4. Yearly maintenance of the community center is figured at $27 per person per year. In all, 247 senior citizens, 56 other adults, 132 young people, and 210 children use the center. What is the center's yearly maintenance costs?

   _____

5. In December, there is a community song festival. The budget for this event is $700. Refreshments cost $437. Costumes for the singers cost $153. Decorations cost $94, and miscellaneous costs come to $45. Did the song festival stay within the budget? _____

6. By December, the community center had raised $93,245 toward its goal of $100,000. A raffle was held to raise the rest of the money. How many $5 raffle tickets have to be sold for the community center to reach its goal?

   _____

Use with pages 90–91.

PRACTICE: Multiplying Decimals by Whole Numbers

Multiply and find the product on the wheel. Write the letter of the answer.

1. 26.4 × 13 _____
2. 21 × 0.233 _____
3. 8.91 × 4 _____
4. 0.6 × 9 _____
5. 52 × 0.047 _____
6. 0.001 × 25 _____
7. 1.073 × 7 _____
8. 2.135 × 31 _____
9. 0.82 × 15 _____
10. 0.0099 × 23 _____

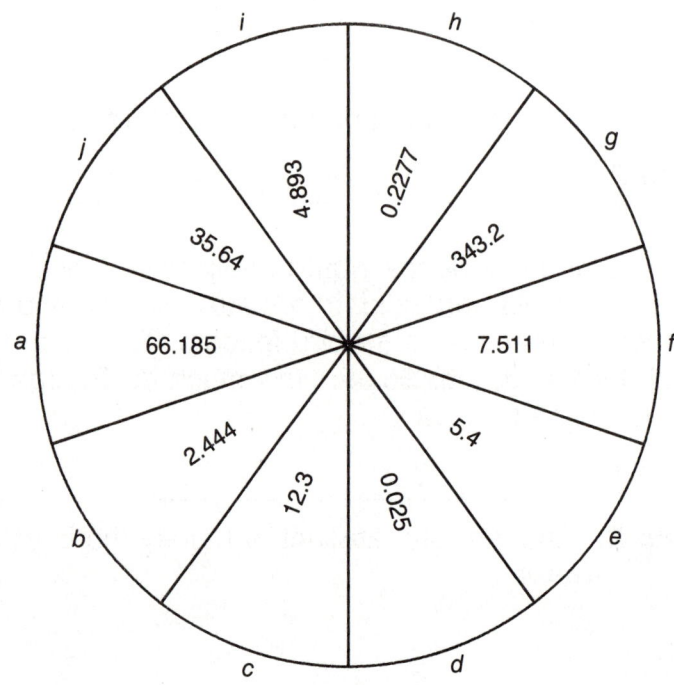

Multiply.

11. 0.51 × 13	12. 4.13 × 3	13. 77.4 × 12	14. 3.51 × 25	15. 0.028 × 8
16. 7.862 × 36	17. 9.157 × 41	18. 5.739 × 4	19. 0.1333 × 13	20. 9.522 × 9
21. 4.45 × 25	22. 0.099 × 19	23. 8.48 × 37	24. 0.848 × 37	25. 0.99 × 19

Use with pages 92–93.

PRACTICE: Estimating Decimal Products

Estimate. Then write > or < to show how you would adjust the estimate.

1. 3.54 × 1.89 = _____
2. 9.14 × 0.11 = _____
3. 0.79 × 8.2 = _____
4. 9.23 × 4.12 = _____
5. 7.92 × 1.98 = _____
6. 7.21 × 0.93 = _____
7. 3.18 × 4.87 = _____
8. 0.76 × 5.1 = _____

Estimate. Write > or < for ◯.

9. 0.63 × 0.427 ◯ 0.2
10. 4.43 × 0.81 ◯ 4.5
11. 7.59 × 0.36 ◯ 2.2
12. 5.06 × 7.32 ◯ 40
13. 0.3792 × 0.9178 ◯ 0.25
14. 15.75 × 0.18 ◯ 3.5

Estimate to find the most sensible answer. Write the letter of the correct answer.

15. 0.27 × 0.369   a. 0.09963   b. 0.9963   c. 9.963   d. 99.63
16. 2.1 × 6.87   a. 0.14427   b. 1.4427   c. 14.427   d. 144.27
17. 8 × 0.83   a. 0.0664   b. 66.4   c. 6.64   d. 0.664
18. 0.41 × 9.26   a. 0.37966   b. 3.7966   c. 37.966   d. 0.03796

Solve.

19. Jerome wants to carpet his room, which measures $10\frac{1}{2}$ ft by $9\frac{3}{4}$ ft. The carpet store has three pieces of carpet that Jerome likes. One piece measures 90 ft², another measures 120 ft², and the third measures 175 ft². Which piece should he choose to carpet his room?

20. Gail works at a local theater during the summers. She is paid $3.85 per hour and works about $6\frac{1}{2}$ hours per day, 3 days per week. About how much does she earn per day?

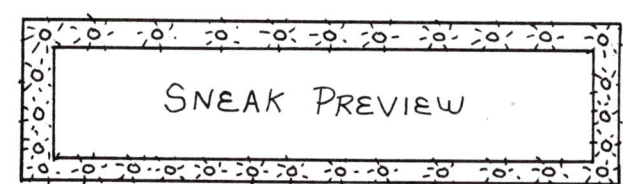

SNEAK PREVIEW

Use with pages 94–95.

PRACTICE: Multiplying by a Decimal

Multiply.

1. 8.895 × 0.36
2. 0.9 × 2.4
3. 46.728 × 0.13
4. 226.08 × 3.1

5. 0.536 × 1.1
6. 27.241 × 0.001
7. 0.85 × 3.29
8. 0.09 × 0.89

9. 67.04 × 0.007
10. 57.382 × 0.41
11. 9,064 × 0.37
12. 0.211 × 0.836

13. 82.9 × 3.05
14. 13.81 × 8.13
15. 109.2 × 0.09
16. 7.35 × 0.23

17. 26.95 × 0.32 = _____
18. 7.4 × 8.1 = _____
19. 473.1 × 0.221 = _____
20. 3.65 × 0.04 = _____
21. 8.9 × 641.3 = _____
22. 9.2 × 78.2 = _____

Multiply. Round to the nearest cent if necessary.

23. $3.75 × 0.11
24. $126.95 × 6.70
25. $4,009.71 × 4.77
26. $0.58 × 0.07

PRACTICE: Guess and Check

Use the guess-and-check method to solve.

1. Everybody on the Fairview Middle School track team is a runner, a jumper, or a lifter depending on whether his or her main event is a running event, a jumping event, or a weight event (such as the shot put). If there are twice as many jumpers as lifters on the team and three times as many runners as lifters and the total number of team members is 30, how many lifters are on the team?

2. Three teams competed in the first meet. The total number of points scored by the three teams was 150. Fairview won the meet by one point; Glenview was second, one point ahead of Appleton. How many points did Fairview score?

3. In a track meet, a competitor earns 5 points for his or her team for taking first place in an event, 3 points for second, and one point for third. Omar scored 11 points for Fairview in the first meet. If he took third place in only one of the three events in which he competed, in how many events did he finish second?

4. Omar's main event is the pole vault. He uses a pole that is 12.8 feet long. When he vaults, the length of pole below the hand that he places on the pole is 3 times the length of pole above that hand. How far up the pole does Omar keep his hand when he vaults?

5. There are 26 people involved with basketball. For every 3 players there is 1 squad leader and for every 6 players there is 1 coach. How many players, squad leaders, and coaches are there?

Use with pages 98–99.

PRACTICE Guessing and Checking

Solve.

6. Fairfield owns 3 kinds of shot puts: junior shot puts, varsity shot puts, and Olympic shot puts. The weight of 4 junior shot puts is the same as the weight of 3 varsity shot puts, and 3 varsity shot puts weigh as much as 2 Olympic shot puts. An Olympic shot put and a junior shot put together weigh 24 pounds. How much does one Olympic shot put weigh?

_____

7. The 440-yard relay is an event in which each of 4 runners carries a short rod called the baton. Each runner carries the baton 110 yards and then passes it to the next runner who continues the race. Fairfield's relay team won this event with a time of 48.0 seconds. If each of the last three Fairfield runners ran his or her 110 yards in 0.4 seconds faster than the preceding runner, what was the time of the runner who finished the race? _____

8. Over the course of the track season, Roger had equal numbers of first-place, second-place, and third-place finishes. A first-place finish is worth 5 points, a second-place finish is worth 3 points, and a third-place finish is worth 1 point. The total number of points Roger scored was 72. How many times did Roger place first, second, and third throughout the entire season? _____

9. Cheryl accumulated 54 points during the season. If the number of times she finished first was twice the number of times she finished second and the number of times she finished second was twice the number of times she finished third, how many times did Cheryl finish third? (Recall that a first-place finish is worth 5 points, a second-place finish is worth 3 points, and a third-place finish is worth one point).

_____

PRACTICE: Division Facts

Write the number that represents each part of the division problem.

$8\overline{)32}$ with quotient 4

$30 \div 5 = 6$

1. divisor _____
2. divisor _____
3. dividend _____
4. dividend _____
5. quotient _____
6. quotient _____

Write the missing number.

7. $12 \div 3 = 4; 4 \times 3 =$ _____
8. $10 \div 2 = 5; 5 \times 2 =$ _____
9. $28 \div 4 = 7; 7 \times 4 =$ _____
10. $56 \div 8 = 7; 7 \times 8 =$ _____
11. $5 \times 5 = 25; 25 \div 5 =$ _____
12. $9 \times 7 = 63; 63 \div 7 =$ _____
13. $17 \times 3 = 51; 51 \div 3 =$ _____
14. $100 \times 6 = 600; 600 \div 100 =$ _____

Divide.

15. $48 \div 6 =$ _____
16. $16 \div 4 =$ _____
17. $49 \div 7 =$ _____
18. $36 \div 9 =$ _____
19. $24 \div 12 =$ _____
20. $81 \div 9 =$ _____
21. $21 \div 3 =$ _____
22. $72 \div 8 =$ _____
23. $25 \div 5 =$ _____

24. $3\overline{)18}$
25. $4\overline{)32}$
26. $6\overline{)24}$
27. $7\overline{)14}$

28. $8\overline{)72}$
29. $5\overline{)25}$
30. $9\overline{)81}$
31. $6\overline{)12}$

Solve.

32. Sue, Mike, and Paul each collected an equal number of bottles. They collected 27 bottles in all. How many bottles did each of them collect?

_____

33. The scout troop collected 48 pounds of newspaper for recycling. Each scout collected 6 pounds of paper. How many scouts are there in the troop?

_____

Use with pages 110–111.

PRACTICE Order of Operations

Compute.

1. $6 \div 3 + 2 =$ _____
2. $2 \div 1 + 3 =$ _____
3. $2 \times 6 - 3 =$ _____
4. $2 \times 2 + 4 =$ _____
5. $8 + 6 \times 9 =$ _____
6. $7 \times 4 - 13 =$ _____
7. $10 \div 5 + 5 =$ _____
8. $(5 + 11) \div 4 =$ _____
9. $(48 - 4) \div 4 =$ _____
10. $12 + 40 \div 2 =$ _____
11. $12 \times (9 - 7) =$ _____
12. $(77 - 65) \div 2 =$ _____
13. $4 \times 7 \div 14 =$ _____
14. $(18 \div 6) \times 5 =$ _____
15. $75 \div 5 \times 3 =$ _____
16. $15 - (3 - 2) =$ _____
17. $4 \times (13 - 1) =$ _____
18. $58 - (25 - 24) =$ _____
19. $(60 \div 6) \times 2 =$ _____
20. $(1 + 50) \div 17 =$ _____
21. $22 \times 4 \div 8 =$ _____
22. $6 \times 5 + 5 =$ _____
23. $19 - (7 + 6) =$ _____
24. $19 - 11 + 11 =$ _____
25. $12 + 3 \times 2 =$ _____
26. $13 + 2 - 5 =$ _____
27. $(36 \div 4) \times 8 =$ _____
28. $7 \times (7 - 6) =$ _____
29. $56 + 2 - 17 =$ _____
30. $33 \div 3 + 8 =$ _____

Write *true* or *false*.

31. $42 \div 3 + 3 = 7$ _____
32. $36 \div (3 + 3) = 6$ _____
33. $(2^3) + 16 \div 8 = 3$ _____
34. $7 - (6 - 1) = 0$ _____
35. $10 + 49 \div (7^2) = 17$ _____
36. $7 + 49 \div 7 = 8$ _____
37. $120 \div 6 + 100 = 120$ _____
38. $(10^2) - 20 + 11 = 91$ _____
39. $6 \times 9 - 54 = 0$ _____
40. $(45 \times 2) \times 2 = 90$ _____
41. $(105 \div 5) \times 9 = 189$ _____
42. $5 \times 25 \div 125 = 1$ _____
43. $(11^2) \div 1 + 10 = 11$ _____
44. $45 \times (9 \div 3) = 270$ _____

PRACTICE: Dividing with Remainders/Divisibility

Divide. Check your answers.

1. 5)29   2. 3)17   3. 8)68   4. 4)39   5. 7)61

6. $\frac{74}{9}$ = _____   7. $\frac{29}{6}$ = _____   8. $\frac{51}{9}$ = _____   9. $\frac{34}{4}$ = _____   10. $\frac{19}{3}$ = _____

Copy the circles and labels below. Write the letter of the number in the correct circle. (A letter may be written in more than one circle.)

a. 25	b. 16	c. 4	d. 90	e. 45
f. 10	g. 250	h. 81	i. 36	j. 522
k. 7,452	l. 5,112	m. 192	n. 5,120	o. 315
p. 85	q. 567	r. 1,498	s. 381	t. 129
u. 640	v. 812	w. 95	x. 246	y. 288

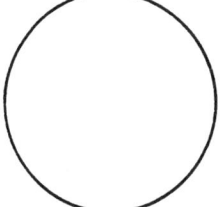
Divisible by 2

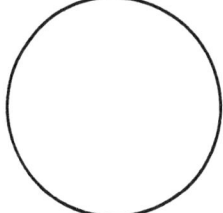
Divisible by 3

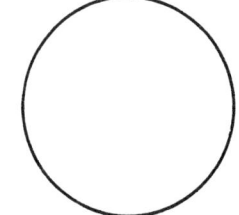
Divisible by 5

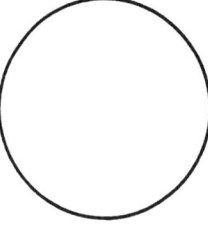
Divisible by 6

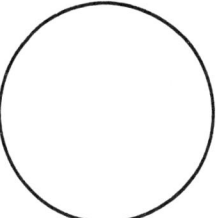
Divisible by 9

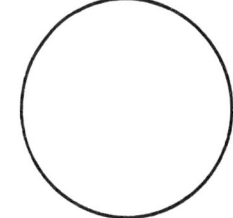
Divisible by 10

John, the winner of the City Marathon, has been invited to City Hall to meet the mayor. In order to reach City Hall, he can only use street numbers that are divisible by 8. List the streets John can use.

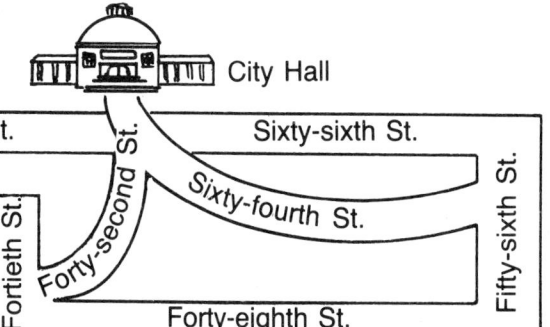
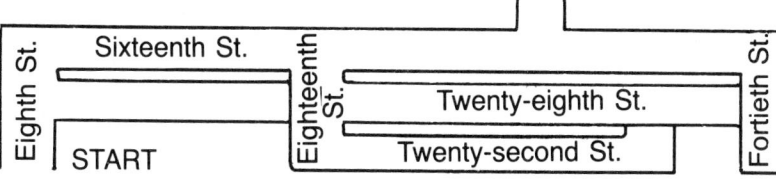

Use with pages 114–115.

PRACTICE: Estimating Quotients

Match each division problem in Column 1 to a range of numbers in Column 2 that contains a good estimate of the quotient. Copy the answer rules and numbers below. Then transfer each letter to the numbered space below that corresponds to the matched problem number. When you are finished, you will have spelled out the answer to the question: "What did the little acorn say when it grew up?"

Column 1

1. 6)2,520
2. 13)53,695
3. 7)15,428
4. 8)9,017
5. 25)518,603
6. 3)643,219
7. 18)2,160

Column 2

- O 2,000 – 3,000
- Y 100 – 200
- T 20,000 – 30,000
- G 400 – 500
- M 1,000 – 2,000
- R 200,000 – 300,000
- E 4,000 – 5,000

__ __ __ __ __ __ __ __ !
 1   2   3   4   2   5   6   7

Write the letter of the best estimate.

8. 6)7,296	a. 100	b. 1,000	c. 10,000
9. 3)25,841	a. 800	b. 8,000	c. 80,000
10. 12)4,983	a. 400	b. 40	c. 4,000
11. 18)365,893	a. 200	b. 2,000	c. 20,000
12. 9)281,365	a. 30,000	b. 3,000	c. 300
13. 26)5,396	a. 2,000	b. 200	c. 20
14. 7)9,205	a. 10	b. 100	c. 1,000
15. 15)45,271	a. 300	b. 3,000	c. 30,000

PRACTICE: Dividing by a 1-Digit Divisor

Divide. Check by multiplying.

1. 3)39
2. 5)60
3. 2)45
4. 7)$469

5. 7)80
6. 4)$0.24
7. 9)108
8. 1)36

9. 6)359
10. 3)172
11. 8)667
12. 9)$2.52

13. 6)211
14. 4)312
15. 7)203
16. 4)212

17. 424 ÷ 8 = _____
18. 225 ÷ 5 = _____
19. 126 ÷ 2 = _____
20. 238 ÷ 3 = _____
21. 327 ÷ 4 = _____
22. 418 ÷ 6 = _____
23. 300 ÷ 5 = _____
24. 758 ÷ 9 = _____
25. 539 ÷ 7 = _____

Solve.

26. At the Annual Woodland Volleyball Tournament, 342 athletes played. How many teams of 6 players participated?

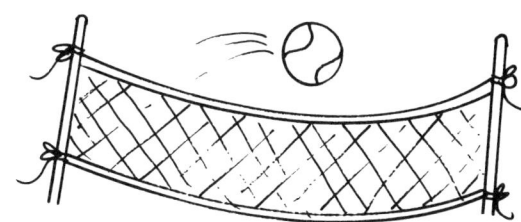

Use with pages 118–119.

PRACTICE: Dividing Larger Numbers

Divide. Check by multiplying.

1. 5)2,740
2. 6)$39.12
3. 8)4,401
4. 9)8,380

5. 4)$38.56
6. 7)6,249
7. 4)9,023
8. 3)2,403

9. 8)523
10. 6)8,405
11. 2)4,039
12. 7)6,413

13. 7)$497
14. 3)2,792
15. 5)6,008
16. 4)3,143

17. 2,960 ÷ 8 = _____
18. $57.84 ÷ 3 = _____
19. 1,999 ÷ 9 = _____
20. 4,494 ÷ 5 = _____
21. 3,304 ÷ 6 = _____
22. 2,809 ÷ 7 = _____
23. 471 ÷ 6 = _____
24. 8,931 ÷ 5 = _____
25. 258 ÷ 8 = _____
26. 5,648 ÷ 9 = _____
27. 3,947 ÷ 7 = _____
28. 3,825 ÷ 4 = _____

PRACTICE Problem Solving: Practice

Write the letter of the most reasonable estimate.

1. Ella runs a cider press. She pays an average of $0.75 per bushel for apples. How much would she pay for 19,000 bushels?

   a. $14,000   b. $25,000   c. $8,000

2. If it takes 3 bushels of apples to make 2 gallons of cider, how many gallons can Ella produce using 14,000 bushels of apples?

   a. 10,000   b. 14,000   c. 8,000

Write a plan to solve each problem.

3. It takes 3 bushels of apples to make 2 gallons of cider. Ella pays $0.75 per bushel. How much will she charge per gallon if she wants to make $0.50 profit on each bushel she buys?

   Step 1: _____

   Step 2: _____

4. Ella sold 331 gallons of cider on the first Saturday of November. She sold 279 half gallons. Gallons cost $2.72 and half gallons cost $1.60. How much money did Ella take in that day?

   Step 1: _____

   Step 2: _____

Write the letter of the correct equation.

5. Ella uses 2.2 times more Cortland apples than McIntosh apples. If she buys 2,208 bushels of McIntosh apples, how many bushels of Cortlands does she buy?

   a. $2{,}208 = n \times 2.2$

   b. $2.2 \times 2{,}208 = n$

   c. $2{,}208 \times n = 2.2$

6. Ella buys 1,104 bushels from Red Apple Farms. She buys 1,178 more bushels from Applegate Orchards. How many bushels does she buy from Applegate Orchards?

   a. $1{,}104 + 1{,}178 = n$

   b. $n + 1{,}178 = 1{,}104$

   c. $1{,}104 - 1{,}178 = n$

Write the letter of the correct operation.

7. Ella buys an average of 880 bushels of apples per month. How many bushels does she buy in a year?

   a. addition

   b. subtraction

   c. multiplication

8. Ella earned $231 more during the second week of August than the first. She earned $844 in the second week. How much did she earn during the first week?

   a. addition

   b. subtraction

   c. multiplication

Use with pages 122–123.

PRACTICE — Problem Solving Practice

Solve.

9. Stuart runs a printing press. His rent is $525 per month. In April, he earned $1,149. He spent $139 for materials. What was his profit? _____

10. Stuart plans to print about 210 posters and sell them for between $2 and $4 each. About how much money will this project take in? _____

11. Stuart takes an order for 250 invitations. Each invitation will cost him 1¢ to print. He will charge 5¢ per invitation. How much money will he earn? _____

12. The printing press was built 81 years before Stuart was born. Stuart was born in 1956. In what year was the printing press built? _____

13. Stuart prints posters at a rate of 12 per minute. He has an order that will take him 30 minutes to print. How many posters is this order for? _____

14. The printing press uses lead type which is stored in drawers. There are 44 drawers of type. Each drawer holds about 2,500 letters. About how many letters are there in all of the drawers? _____

15. Stuart prints a small magazine. The magazine uses 24 sheets of paper per issue. He prints 200 copies of each issue. There have been 6 issues of the magazine. What is the total number of sheets of paper used in all of the issues printed? _____

16. Stuart can print 26 invitations per minute. He has one order that will take him 12.5 minutes to print. He has a second order that will take him twice as long as the first. How many invitations will he print for the second order? _____

17. Stuart is working on a poster for an art fair. It will take him 1 hour and 40 minutes to set up the lead type, half as long to ink the press, and 2 hours to print copies of the poster. How much longer will he spend printing the poster than inking the press? _____

18. Stuart is planning to print a book. The book will be between 180 and 200 pages long. He plans to print about 435 copies. He wants to order paper on which to print the book. How many pages should he order to be sure he has enough to print the book? _____

PRACTICE Short Division

Use short division. Check your answer.

1. 2)1,259
2. 4)96
3. 7)458
4. 6)750

5. 5)615
6. 2)687
7. 9)5,067
8. 5)1,297

9. 3)7,383
10. 5)1,230
11. 8)831
12. 2)2,390

13. 4)225
14. 6)1,410
15. 3)372
16. 3)788

17. 4)795
18. 5)2,735
19. 6)4,162
20. 3)183

21. 7)1,141
22. 6)264
23. 8)6,210
24. 3)739

25. 2)4,736
26. 9)581
27. 3)2,164
28. 5)6,255

29. 5)$6.65
30. 7)$30.24
31. 4)$65.68
32. 3)$7.89

Solve.

33. Walter has 2,544 pamphlets to pass out at 4 schools. He wants each school to have the same number of pamphlets. How many pamphlets should he give to each school? _____

34. Walter has 6 more sets of pamphlets printed. The bill for the printing comes to $97.86. How much does it cost to print one set of pamphlets? _____

Use with pages 124–125.

PRACTICE: Estimating

Write *underestimate* or *overestimate* to complete each sentence.

UNITED STATES CAR SALES

Monday	18,567	Thursday	17,918
Tuesday	20,189	Friday	21,568
Wednesday	25,632	Saturday	27,811

1. Automakers try to meet a projected sales figure each week. This figure is based on sales for the same week during the previous year. Automakers should _____ when trying to determine whether they will reach the projected figure.

2. Automakers need to order parts in order to assemble cars in one location. To ensure that enough parts are ordered according to the chart above, they should _____ the number of cars that will be sold in each of the coming weeks.

Use estimated amounts to solve the problems. Use exact amounts only if you need to.

CAR PRICE LIST

Fury	$8,599
Montreal	$11,179
Charger	$8,459
Coupe d'Elegance	$12,139
Arrow	$6,950
Carrington	$11,690

3. A Fury dealer can spend $36,000 this week for new Furies for his inventory. Does the dealer have enough money to buy 4 cars? _____

4. If the leasing company uses $48,250 to buy one of each of the first 5 cars on the list, how much more money do they need to also buy the Carrington? _____

5. A leasing company wants to buy 5 cars to use for rental purposes. They have $43,250 to spend on 5 cars. Does the company have enough money to buy 5 cars from the list? _____

6. A Montreal dealer stocks both Furies and Montreals. The dealer has $42,500 to spend for his inventory. If the dealer buys 2 Furies and 2 Montreals, how much money will he have left? _____

PRACTICE Dividing by Multiples of 10

Divide.

1. 360 ÷ 60 = _____
2. 560 ÷ 80 = _____
3. 2,100 ÷ 30 = _____
4. 400 ÷ 50 = _____
5. 3,600 ÷ 40 = _____
6. 63,000 ÷ 90 = _____
7. 250 ÷ 5 = _____
8. 120 ÷ 6 = _____
9. 4,900 ÷ 700 = _____

10. 20)800
11. 9)720
12. 30)240
13. 60)540

14. 40)1,600
15. 70)3,500
16. 50)30,000
17. 90)45,000

18. 30)18,000
19. 20)14,000
20. 60)4,200
21. 40)4,000

22. 50)5,500
23. 700)3,500
24. 800)48,000
25. 30)36,000

Solve.

26. James takes the bus to school every morning with his friends. The school provides 30 buses for its 780 students, and each bus carries exactly the same number of students. How many students travel to school in James's bus?

27. The school bus can travel 440 miles on 20 gallons of gasoline. What is the average number of miles per gallon that the bus gets?

Use with pages 128–129.

PRACTICE: Dividing by a 2-Digit Divisor

Divide. Check your answer.

1. 12)48
2. 23)46
3. 50)150
4. 17)56

5. 25)759
6. 85)340
7. 77)291
8. 64)389

9. 32)128
10. 90)811
11. 14)615
12. 40)239

13. 15)345
14. 23)631
15. 36)278
16. 11)121

17. 308 ÷ 43 = _____
18. 416 ÷ 52 = _____
19. 275 ÷ 55 = _____
20. 249 ÷ 17 = _____
21. 430 ÷ 35 = _____
22. 135 ÷ 13 = _____
23. 216 ÷ 22 = _____
24. 756 ÷ 28 = _____
25. 490 ÷ 15 = _____
26. 952 ÷ 19 = _____
27. 549 ÷ 16 = _____
28. 64 ÷ 15 = _____
29. 584 ÷ 11 = _____
30. 330 ÷ 88 = _____
31. 871 ÷ 13 = _____

PRACTICE: Correcting Estimates

Divide.

1. 56)392
2. 41)246
3. 17)731
4. 23)184
5. 32)128

6. 4)139
7. 15)950
8. 32)214
9. 16)184
10. 25)90

11. 38)956
12. 37)296
13. 47)724
14. 35)695
15. 23)667

16. 32)624
17. 24)816
18. 16)438
19. 21)594
20. 43)731

21. $\frac{630}{76} =$ _____
22. $\frac{196}{24} =$ _____
23. $\frac{398}{66} =$ _____
24. $\frac{570}{31} =$ _____
25. $\frac{515}{53} =$ _____

Solve.

26. A shipper has 156 grapefruits. Each crate can hold 52 grapefruits. How many crates can the shipper fill?

27. It takes 48 oranges to make a gallon of freshly squeezed juice. How many gallons of juice can be made with 672 oranges?

Use with pages 132–133.

PRACTICE — Dividing Thousands

Complete each division exercise by matching the key and the correct door.

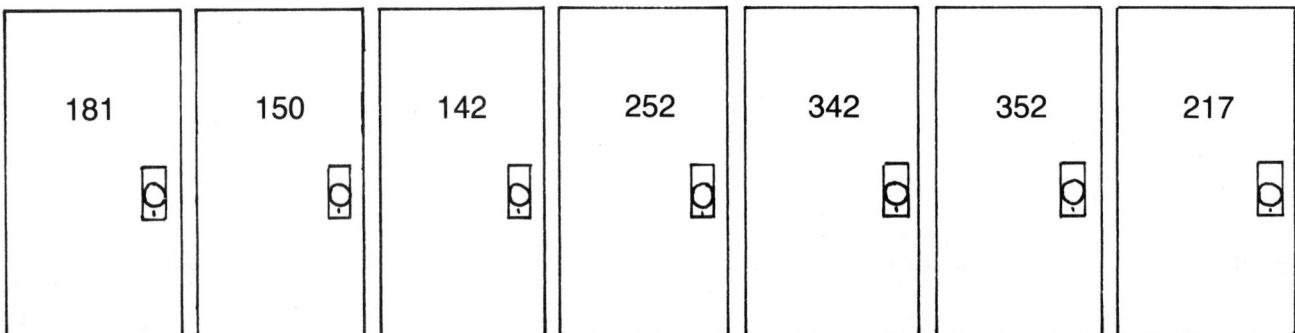

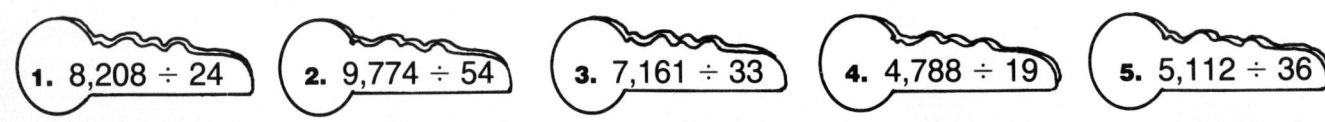

Divide.

6. 48)$583.68

7. 32)1,396

8. 29)3,944

9. 54)648

10. 17)4,099

11. 73)$702.99

12. 37)1,702

13. 69)1,530

14. 26)525

15. 14)$212.38

16. 41)1,298

17. 85)935

PRACTICE Dividing Larger Numbers

Divide and check.

1. 15)9,572
2. 91)453,890
3. 11)$161.15
4. 40)$2,798.00

5. 35)62,005
6. 61)584,746
7. 18)130,572
8. 35)73,955

9. 24)110,479
10. 10)55,750
11. 96)423,710
12. 12)78,780

13. 20)908,420
14. 14)76,045
15. 16)841,873
16. 74)140,452

17. $10,516.50 ÷ 41 = _____
18. $3,706.20 ÷ 58 = _____

19. 250,170 ÷ 17 = _____
20. $35.55 ÷ 15 = _____

Use with pages 136–137.

PRACTICE — Zeros in the Quotient

Divide.

1. 48)9,984

2. 68)1,147

3. 43)2,580

4. 26)6,240

5. 27)21,762

6. 32)6,412

7. 30)22,304

8. 53)34,450

9. 22)8,866

10. 31)62,217

11. 74)7,548

12. 15)45,700

13. 12,409 ÷ 12 = _____

14. 6,479 ÷ 31 = _____

15. 18,400 ÷ 20 = _____

16. 1,215 ÷ 71 = _____

17. 6,617 ÷ 13 = _____

18. 3,564 ÷ 33 = _____

19. 48,312 ÷ 44 = _____

20. 3,105 ÷ 15 = _____

Solve.

21. The chef at Camp Big Sky ordered 2,520 cans of tuna. If there are 24 cans in each case, how many cases were ordered?

22. The chef works for 12 weeks during the summer and earns $3,685. What is the chef's average weekly salary?

PRACTICE: Dividing by a 3-Digit Divisor

Divide and check.

1. 183)833,658   2. 112)570   3. 255)14,823   4. 860)60,225

To solve the riddle, copy the answer rules and numbers below. Divide. Find the remainder for each problem. Then write the letter next to each answer on the line over the remainder.

5. 225)201,975   C   6. 193)53,127   R   7. 334)9,197,800   U

8. 636)663,510   O   9. 114)220,582   N   10. 316)616,653   H

11. 363)885,122   B   12. 271)869,333   A   13. 127)26,493   T

When the apple tree wanted to try something new, what did the pear tree tell it to do?

___ ___ ___ ___ ___ ___
128 52 236 106 150 137

___ ___ ___
162 108 77

Use with pages 140–141.

PRACTICE: Equations

Write the value of n.

1. $49 \div n = 7$
   n = _____

2. $69 \div n = 3$
   n = _____

3. $124 \div n = 31$
   n = _____

4. $324 \div n = 18$
   n = _____

5. $72 \div n = 6$
   n = _____

6. $27 \div n = 3$
   n = _____

7. $12 \div n = 1$
   n = _____

8. $114 \div n = 19$
   n = _____

9. $340 \div n = 4$
   n = _____

10. $891 \div n = 99$
    n = _____

11. $7 \times n = 315$
    n = _____

12. $n \times 11 = 616$
    n = _____

13. $n \times 20 = 280$
    n = _____

14. $n \div 40 = 60$
    n = _____

15. $n \times 5 = 665$
    n = _____

16. The product is 450.
    One factor is 50.
    The other factor is n.
    n = _____

17. The dividend is 200.
    The quotient is 100.
    The divisor is n.
    n = _____

18. The dividend is 52.
    The quotient is 13.
    The divisor is n.
    n = _____

19. The product is 820.
    One factor is 41.
    The other factor is n.
    n = _____

20. The divisor is 6.
    The quotient is n.
    The dividend is 186.
    n = _____

21. The divisor is 18.
    The dividend is 540.
    The quotient is n.
    n = _____

PRACTICE: Interpreting the Quotient and the Remainder

Jeremy, Jessy, Randy, and Rod Ashton raise fruits and vegetables. One morning in September, they are packing their truck with produce to sell in town.

Solve.

Jeremy has a pile of 235 ripe apples. He fills as many bags as he can with 8 apples each.

1. How many apples does Jeremy leave behind? _____

2. How many bags of apples does he fill? _____

Next, Jessy loads squashes into bushel baskets. He has 59 squashes, and each bushel basket will hold 11 squashes.

3. If Jessy fills as many baskets as he can, what is the fewest number of squashes any basket holds? _____

4. How many bushel baskets will Jessy need to take all his squashes to town? _____

Next, Randy stuffs ears of corn into burlap sacks. Each sack will hold 60 ears, and Randy has 513 ears to pack.

5. What is the smallest number of sacks Randy will need in order to get all his corn to market? _____

6. If he fills as many bags as he can full of corn, how many full bags of corn does he bring? _____

Next, Rod packs raspberries into pint containers. He estimates that each container holds 75 raspberries.

7. How many full pint-containers can be packed from a pile of 1,578 raspberries? _____

8. How many containers would he need to pack all the raspberries in a pile of 2,406? _____

The Ashton's truck gets 17 miles per gallon of gas.

9. How many gallons of gas should they buy to go 56 miles? _____

10. The gas tank was empty before they added 6 gallons of gas. How much gas is left after they drive 93 miles? _____

Use with pages 144–145.

PRACTICE: Dividing Decimals by Whole Numbers

Divide.

1. 5)2.5025
2. 30)28.7160
3. 8)7.4496
4. 36)51.5808

5. 3)5.6664
6. 13)32.7041
7. 23)74.0117
8. 5)2.2560

9. 6)8.532
10. 17)0.8721
11. 33)98.9967
12. 8)$9.44

13. 3)2.1303
14. 13)13.3237
15. 2)9.662
16. 11)94.1611

Solve.

17. At a Mexican festival, there were 5 piñatas. The total weight of the gifts in the piñatas was 32.25 pounds. If each piñata had an equal weight of gifts, what was that weight?

_____

18. In Mexico, the Culvers drove 206.64 miles in 4 hours. What was the average distance they drove each hour?

_____

PRACTICE More Dividing by Whole Numbers

Divide.

1. 5)2.66
2. 6)21
3. 8)4
4. 6)0.357

5. 2)0.1
6. 2)41
7. 4)0.002
8. 8)425

9. 5)$6.20
10. 12)16.98
11. 6)$123
12. 5)9

13. 4)$18
14. 24)331.8
15. 18)387
16. 16)20

17. 23.67 ÷ 45 = _____
18. 1.995 ÷ 42 = _____
19. $655.50 ÷ 75 = _____
20. $518.40 ÷ 15 = _____
21. 6.75 ÷ 18 = _____
22. 32 ÷ 64 = _____
23. 36.63 ÷ 66 = _____
24. $25.20 ÷ 35 = _____
25. 130 ÷ 52 = _____
26. 13 ÷ 52 = _____

Use with pages 156–157.

PRACTICE: Dividing Decimals by 10; 100; or 1,000

Divide by 10.

1. 46.1 _____
2. 309.72 _____
3. 88.4 _____
4. 723.4 _____

5. 60.24 _____
6. 451.6 _____
7. 100.01 _____
8. $11.90 _____

Divide by 100.

9. 58.2 _____
10. $94.00 _____
11. $617.00 _____
12. 359.82 _____

13. 75.8 _____
14. 197.65 _____
15. 768.99 _____
16. 45.49 _____

Divide by 1,000.

17. 43,004.1 _____
18. $760.00 _____
19. 823.59 _____
20. $4,160.00 _____

21. 7,598.2 _____
22. $110.00 _____
23. 606.09 _____
24. $540.00 _____

Divide.

25. 526 ÷ 100 = _____
26. 72.3 ÷ 10 = _____
27. 514.88 ÷ 1,000 = _____

28. 10)72.3
29. 10)$64.50
30. 1,000)8,490.00

Solve.

31. The concession owner at Stell Stadium wants to know how much food per person is sold at baseball games. On Sunday 1,000 people attended a baseball game and consumed 839 hot dogs and 1,026 bags of peanuts. How many hot dogs were sold per person? _____

How many bags of peanuts were sold per person?

_____

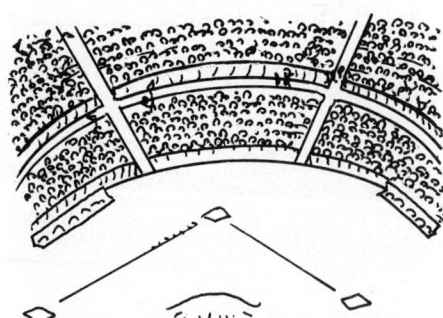

PRACTICE: Dividing by a Decimal

Divide.

1. 5.2)$16.12

2. 0.2)0.1234

3. 0.055)0.0220

4. 30.6)247.86

5. 0.09)0.684

6. 8.7)0.0783

7. 0.3)0.99999

8. 5.4)0.162

9. 3.17)1.902

10. 0.006)0.024

11. 7.8)$157.56

12. 0.5)$37.55

13. 0.2)0.22222

14. 8)0.392

15. 4.4)4.312

16. 0.5)0.52525

17. 2.5)21.03

18. 0.87)1.131

19. 5)213.55

20. 9.1)2.2386

21. 352.03 ÷ 74.9 = _____

22. 0.224 ÷ 0.056 = _____

23. 0.527 ÷ 0.062 = _____

24. 0.609 ÷ 0.087 = _____

25. $17.49 ÷ $3.30 = _____

26. 6.24 ÷ 2.6 = _____

Use with pages 160–161.

PRACTICE: More Dividing by a Decimal

To solve the riddle, copy the answer rules and numbers below. Divide. Find your answer and write the letter next to the problem above the line.

1. 0.8)2.88   T
2. 7)0.343   R
3. 2.3)15.18   K
4. 0.32)0.464   H

5. 0.9)3.483   A
6. 15)$9.45   I
7. 3.4)293.42   M
8. 2.15)0.5805   C

9. 0.8)10.08   T
10. 0.015)0.0585   E
11. 61)$15.25   R
12. 0.5)244.50   S

What do mathematical acrobats do?

___ ___ ___ ___ ___ ___ ___ ___ ___ ___ ___ ___
3.87 0.049 $0.63 3.6 1.45 86.3 3.9 12.6 $0.25 $0.63 0.27 6.6 489

PRACTICE Choosing the Operation

All 187 students in the Southfield sixth grade take a trip to the Yellowjacket Copper Mine.

Write which operations are required. Solve.

1. The class breaks up into groups, each of which contains no more than 7 students. How many students are in the group with less than 7?

2. If each student in a group of 7 brings back 3.5 kilograms of mineral specimens, what is the total weight of ore collected by the group?

3. Jane Fleming's group picked up 4.35 kilograms of deep blue azurite, 3.67 kilograms of bright green malachite, 1.69 kilograms of brassy chalcopyrite, 0.31 kilograms of quartz crystals, and 5.55 kilograms of unidentified specimens. What was the weight of all they collected?

4. Jerry Sykes' group brought back 13.63 kilograms more than Dianne Luzac's group, which collected 2.57 kilograms less than Mark Jacobi's group. If the total weight of ore collected by Mark's group was 9.67 kilograms, what was the weight of ore collected by Jerry's group?

A sample of 1,000 grams of ore taken from the main pit of the mine is found to contain 5.578 grams of copper.

5. What was the weight of material other than copper in the sample taken?

6. On the average, what was the weight of copper in a single gram of the sample taken?

7. About how many times greater was the weight of the ore sample taken than the weight of the copper it contained?

8. A molecule of azurite contains 3 atoms of copper, 8 atoms of oxygen, 2 of carbon, and 2 of hydrogen. How many more atoms of oxygen than copper are contained in 6,357 molecules of this mineral?

Use with pages 164–165.

PRACTICE: Rounding Decimal Quotients

Divide. Round to the nearest tenth.

1. 3)8
2. 74)100
3. 0.96)52
4. 12)154
5. 17)168

Divide. Round to the nearest hundredth or to the nearest cent.

6. 6)$94.21
7. 2.8)73.07
8. 8)$0.95
9. 0.06)0.25
10. 0.4)671
11. 0.063)0.4588
12. 59)$374.11
13. 29.2)45.86
14. 9)79
15. 6.3)16.378

Divide. Round to the nearest dollar.

16. 4.06)$93.65
17. 3)$101.20
18. 45)$446.06
19. 25)$125.50
20. 2.4)$291.24

PRACTICE: Writing an Equation

Write the letter of the correct equation to solve each problem.

1. Bob's farm has 1.4 times as many chickens as cows. The farm has 80 cows. How many chickens does the farm have?
   a. $80 \div 1.4 = n$
   b. $80 \times 1.4 = n$
   c. $1.4 \times n = 80$

2. The farm's chickens lay an average of 36 eggs per day. How many days will it take the chickens to lay 324 eggs?
   a. $324 \div 36 = n$
   b. $324 \times n = 36$
   c. $36 \div 324 = n$

Write an equation. Solve.

3. The average weight of an egg is 1.2 ounces. How much will a carton of a dozen eggs probably weigh?

   _____

4. The farm has $\frac{1}{3}$ as many horses as hogs. The total number of hogs is 51. How many horses does the farm have?

   _____

5. Bob plants 115 rows of corn, 260 rows of rye and 299 rows of wheat. How many times more wheat does he plant than corn?

   _____

6. Bob is transplanting corn plants. Each plant uses 2.3 square feet of ground. How many square feet are needed to plant 221 corn plants?

   _____

7. Bob bought a 21–acre plot of land, which he divided equally among 5 crops. How many acres of land is reserved for each crop?

   _____

8. Bob buys chicken feed in 30–pound sacks. Each sack lasts 40 days. How much chicken feed does the farm use per day?

   _____

9. A hog on Bob's farm gains 1.22 pounds per day. How many pounds does it gain in a week?

   _____

10. The chicken coop on the farm has an area of 515 square feet. The hog pen's area is 1,607 square feet. How much larger is the hog pen than the chicken coop?

    _____

Use with pages 168–169.

PRACTICE Metric Units of Length

Choose the correct measurement.

1.	length of a small paper clip	33 mm	33 dm	33 m
2.	length of a pickup truck	5 mm	5 m	5 km
3.	length of your little finger	5 cm	5 dm	5 m
4.	length of a pencil	120 m	120 km	120 mm
5.	length of your shoe	24 cm	24 dm	24 mm
6.	distance you ride in a train	50 mm	50 cm	50 km

Which unit would you use to measure? Write *cm, m,* or *km*.

7. a car trip _____   8. your classroom _____   9. a meatloaf _____

10. an oak tree _____   11. an elephant _____   12. a stapler _____

13. a postcard _____   14. hands of a clock _____   15. the Statue of Liberty _____

Measure to the nearest centimeter.

16. _____ _____ cm

17. _____ _____ cm

18. _____ _____ cm

Measure to the nearest millimeter.

19. _____ _____ mm

20. _____ _____ mm

21. _____ _____ mm

22. _____ _____ mm

23. _____ _____ mm

PRACTICE: Equal Metric Measures of Length

km kilometer 1 km = 1,000 m	hm hectometer 1 hm = 100 m	dam dekameter 1 dam = 10 m	m meter	dm decimeter 10 dm = 1 m	cm centimeter 100 cm = 1 m	mm millimeter 1,000 mm = 1 m

Complete.

1. 8 cm = _____ mm
2. 16 mm = _____ m
3. 25 dm = _____ cm
4. 8 cm = _____ dm
5. 42 cm = _____ km
6. 9 dam = _____ cm
7. 7 m = _____ dam
8. 9 dam = _____ km
9. 16 mm = _____ dam
10. 42 cm = _____ mm
11. 1.600 mm = _____ km
12. 23 hm = _____ m
13. 5 m = _____ hm
14. 30 cm = _____ m
15. 5 m = _____ km
16. 25 dm = _____ m
17. 4 dm = _____ cm
18. 6 hm = _____ cm
19. 9 dam = _____ m
20. 3 km = _____ m
21. 182.5 m = _____ cm
22. 25 dm = _____ dam
23. 6 hm = _____ km
24. 42 cm = _____ km
25. 16 mm = _____ cm
26. 7 dam = _____ km
27. 88 cm = _____ mm
28. 6.7 cm = _____ m
29. 16 mm = _____ dm
30. 5 m = _____ dam

31. A desk that is 2,250 mm long = _____ m.
32. A fence that is 270 cm long = _____ m.
33. A dog that is 675 mm long = _____ cm.
34. A stretch of highway that is 3,200 m long = _____ km.

Use with pages 172–173.

PRACTICE Problem Solving: Practice

Write the letter of the correct operation.

1. A paper plane called Skydevil won a flying competition with a flight of 25.20 meters. Skydevil is 0.12 meters long. How many times its own length did Skydevil soar on its winning flight?

   a. addition
   b. division
   c. multiplication
   d. subtraction

2. Each plane in the competition is allowed 4 flights. A plane called Swooper had flights of 21.11 m, 23.53 m, 5.12 m, and 22.70 m. What was the total length of all four flights?

   a. addition
   b. subtraction
   c. multiplication
   d. division

Write the equation you would use to solve each problem.

3. A plane called Hotwings came in second with a flight of 24.60 meters. The flight took 8.20 seconds. On the average, how many meters per second did Hotwings travel?

   _____

4. The Swooper's longest flight was 23.53 m. Its shortest flight was a nosedive that covered only 5.12 m. How much farther did Swooper travel on its longest flight than on its shortest?

   _____

Write the letter of the best estimate.

5. A plane called Fly Tiger had flight times of 2.5 seconds, 4.4 seconds, and 8.1 seconds. Is the time of its second flight closer to the time of its first or third?

   a. closer to first
   b. closer to third
   c. can't answer

6. The Fly Tiger's average distance is 20.12 m. About how far will Fly Tiger travel in 12 flights?

   a. 200 m
   b. 240 m
   c. 350 m

Write the letter of the method you would use to solve.

7. Kathy makes 1 paper airplane from 96 square inches of paper. How many airplanes can she make from 391 square inches of paper?

   a. round quotient up
   b. round quotient down
   c. use remainder only

8. Glen makes one paper plane from 24 square inches of paper. If he makes as many planes as he can from 275 square inches of paper, how much paper will be left?

   a. round quotient up
   b. round quotient down
   c. use remainder only

PRACTICE Problem Solving: Practice

Solve. Use the chart below to provide needed information.

RESULTS OF THE MAPLEWOOD PAPER-PLANE COMPETITION

Name of Plane	Longest Flight	Shortest Flight	Longest Time	Shortest Time
Sky Devil	25.20 m	18.80 m	7.6 sec	7.1 sec
Hotwings	24.60 m	3.22 m	8.2 sec	1.4 sec
Swooper	23.53 m	5.12 m	7.8 sec	1.6 sec
Fly Tiger	21.11 m	20.01 m	8.1 sec	5.5 sec
Air-Gator	19.00 m	12.82 m	9.5 sec	4.0 sec
Spacer	17.75 m	5.95 m	7.7 sec	1.7 sec

9. Spacer's shortest flight also took its shortest time. How many meters per second was Spacer traveling on its shortest flight? _____

10. Fly Tiger's second-longest flight was 0.05 meters shorter than its longest flight. How long was Fly Tiger's second-longest flight? _____

11. Air-Gator's longest flight was also its longest time. Its second-longest flight traveled at the same rate as the longest flight, but for 1.5 seconds less time. How long was Air-Gator's second-longest flight? _____

12. Which plane had the largest variation between the distances of its longest and shortest flights? Which plane had the least variation between the speeds of its longest and shortest flights? _____

13. Any flight under 6 meters is considered to be a nosedive. What is the total distance of nosedives shown on this chart? _____

14. Emily has designed a plane made from balsa wood that can fly 5 times farther than Swooper's longest distance. How far can Emily's plane fly? _____

15. Emily can make one balsa-wood plane from an 18 in. piece of wood. How many planes can she make from a 97 in. piece of wood? _____

16. Ryan makes 2 sizes of planes out of balsa wood. One size uses 19.5 in. of wood, the other uses 12.2 in. of wood. If Ryan makes as many of the larger planes as he can from a 91.2 in. piece of wood, will there be enough wood left to make one of the smaller planes? _____

Use with pages 174–175.

PRACTICE: Metric Units of Capacity and Mass

1 L = 1,000 mL
1 kL = 1,000 L

1 g = 1,000 mg
1 kg = 1,000 g

Complete.

1. 1 mL = _____ L
2. 125 mg = _____ g
3. 10 mL = _____ L
4. 125 mg = _____ kg
5. 100 mL = _____ L
6. 525 g = _____ mg
7. 5 L = _____ mL
8. 525 g = _____ kg
9. 50 L = _____ mL
10. 7 kg = _____ g
11. 67 mL = _____ L
12. 4 L = _____ mL
13. 83 g = _____ mg
14. 260 mg = _____ kg
15. 0.9 mg = _____ g
16. 39 kg = _____ g
17. 1,633 g = _____ kg
18. 8,750 mL = _____ kL
19. 32 mg = _____ g
20. 9.4 g = _____ mg

Write the better measurement.

21. capacity of a bathtub	20 mL	2,000 mL	20 L
22. capacity of a glass	200 mL	2,000 mL	200 L
23. capacity of an eyedropper	2 mL	2,000 mL	2 L
24. mass of a grain of salt	5 mg	5 g	5 kg
25. mass of a pair of shoes	2 mg	2 g	2 kg

Solve.

26. A jug of apple juice is 0.0025 L. What is the capacity of the jug in milliliters?

27. A large egg has a mass of about 60 grams. What is the mass of a dozen eggs? Answer using kilograms.

PRACTICE: Using a Map

Grace and Tom Flanders spent part of their summer vacation hiking in the area shown on the map at the right.

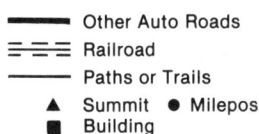

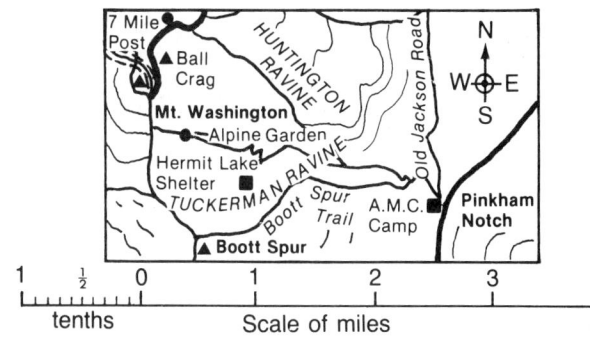

Write the letter of the correct answer.

1. Starting from the Appalachian Mountain Club (A.M.C.) camp in Pinkham Notch, they followed the Boott Spur Trail up to Boott Spur, where they stopped for lunch. How many miles did they hike before lunch?

   a. 1.1 miles   b. 2.0 miles
   c. 3.2 miles

2. It took Grace and Tom 4 h to travel from the A.M.C. camp to a spot 1.5 mi past Boott Spur. How many miles had Grace and Tom traveled per hour?

   a. 1.25 mph   b. 1.5 mph
   c. 0.875 mph

Solve.

3. At Boott Spur, Grace and Tom met Craig, who had walked from a spot 2.4 miles north of Mount Washington. How far had Craig hiked to get to Boott Spur? _____

4. Craig told them about a hike he took last spring. He started at Boott Spur, hiked to the A.M.C. camp, then hiked along a path that took him to Mount Washington by way of the Tuckerman Ravine. How long was this hike? _____

5. Grace and Tom traveled from Boott Spur to Mt. Washington in $2\frac{1}{2}$ hours. How many miles per hour did they travel during this section of their hike? _____

6. Grace and Tom took the same path from Mt. Washington back to the A.M.C. camp. They averaged a rate of 1.2 miles per hour on the way back. How many hours did the return hike take? _____

7. Back at the club, Grace and Tom met Lydia, who had just returned from a hike to Ball Crag. She passed Hermit Lake Shelter and Alpine Garden on her way to the Crag, and the Huntington Ravine on her way back. How long was Lydia's hike? _____

8. Grace and Tom decide to hike from A.M.C. to Alpine Gardens the next day. They plan to hike at a rate of 0.8 miles per hour. They plan to take the most direct path there and back. How much time will this hike probably take? _____

Use with pages 178–179.

PRACTICE Least Common Multiple

Write the letter of the correct answer or answers.

1. Which numbers are multiples of 13?

   a. 52   b. 88   c. 182   d. 690

2. Which numbers are multiples of 15?

   a. 50   b. 90   c. 315   d. 400

3. Which numbers are multiples of 12?

   a. 108   b. 82   c. 312   d. 744

4. Which numbers are multiples of 11?

   a. 888   b. 770   c. 293   d. 671

Solve.

5. List the first six multiples of 20. _____

6. List the first six multiples of 25. _____

7. Write the least common multiple of 20 and 25. _____

8. Write the least common multiple of 120 and 80. _____

9. List the multiples of 35 and 45 until you reach the least common multiple.

   Multiples of 35: _____

   Multiples of 45: _____

Write the least common multiple.

10. 2, 3 _____   11. 2, 5, 7 _____   12. 5, 12, 15 _____

13. 20, 22, 55 _____   14. 90, 70, 30 _____   15. 17, 51 _____

16. 22, 55 _____   17. 150, 200, 600 _____   18. 15, 75, 50 _____

List the multiples of 60 and 75 until you reach the least common multiple.

19. Multiples of 60: _____

20. Multiples of 75: _____

PRACTICE Greatest Common Factor

Write the letter of the correct answer or answers.

1. Which numbers are factors of 300?
   a. 60   b. 45   c. 24   d. 75

2. Which numbers are factors of 90?
   a. 15   b. 18   c. 25   d. 35

List the factors of the number.

3. 10 _____
4. 12 _____
5. 25 _____
6. 26 _____
7. 32 _____
8. 30 _____
9. 60 _____
10. 90 _____

List the common factors.

11. 9, 36 _____
12. 15, 60 _____
13. 48, 64 _____
14. 60, 90 _____
15. 24, 80 _____
16. 18, 36 _____

List the common factors. Ring the greatest common factor.

17. 20, 30, 40 _____
18. 13, 14, 15 _____
19. 25, 50, 75 _____
20. 26, 39, 91 _____
21. 42, 70, 140 _____
22. 84, 60, 96 _____
23. 36, 40, 80 _____
24. 16, 64, 32 _____
25. 32, 48, 72 _____
26. 9, 30, 60 _____
27. 60, 70, 80 _____
28. 36, 45, 81 _____
29. 20, 30, 5 _____
30. 24, 6, 4, 124 _____

Use with pages 192–193.

PRACTICE: Primes, Composites, and Prime Factors

Write the factors of each number. Then write whether the number is *prime* or *composite*.

1. 2 _____
2. 11 _____
3. 8 _____
4. 26 _____
5. 85 _____
6. 61 _____
7. 3 _____
8. 27 _____
9. 83 _____
10. 9 _____
11. 13 _____
12. 31 _____
13. 25 _____
14. 29 _____
15. 39 _____

Copy and complete each factor tree.

16. 40
    4 × 10
    × × ×

17. 100
    4 × 25
    × × ×

18. 27
    × 9
    × ×

19. 90
    15 ×
    × × ×

20. 52
    26 ×
    × ×

21. 30
    ×
    × ×

Write the factors of each number. Ring the prime factors.

22. 70 _____
23. 19 _____
24. 44 _____
25. 75 _____
26. 68 _____
27. 82 _____
28. 24 _____
29. 38 _____

PRACTICE: Writing a Simpler Problem

Write the letter of the better plan for simplifying each problem.

1. Frank works for a produce distributor. On Monday, he delivered 24 crates of lettuce at $26.99 per crate. Tuesday was a very busy day, and Frank delivered twice as many crates of lettuce as he did on Monday. How much money did Frank receive on Tuesday?

   a. Step 1: $2 \times 24 = 48$
   Step 2: $\$26 \times 48 = \$1,248$

   b. Step 1: $2 \times 24 = 48$
   Step 2: $48 + 26 = 74$

2. Frank delivered produce 2 hours and 15 minutes longer on Tuesday than he did on Monday. His deliveries Monday morning took 3 hours and 30 minutes, and the afternoon deliveries took 4 hours and 20 minutes. How long did it take Frank to deliver produce on Tuesday?

   a. Step 1: $3 + 4 = 7$
   Step 2: $7 + 2 = 9$

   b. Step 1: $3 + 2 = 5$
   Step 2: $5 + (4 + 2) = 11$

Solve. Simplify the problem if you need to.

3. Frank is paid $260 a week for delivering produce. If he works more than 8 hours per day, he is paid $9.75 per hour for each hour over 8. What is Frank's weekly pay if he worked 9 hours 3 days this week?

4. Frank's company makes a 15.22% profit on a crate of fruit and a 12.74% profit on a crate of vegetables. What is the company's average-percent profit per crate of produce?

5. The profit per crate of lettuce is $6.25. The profit per crate of cauliflower is $8.35. If Frank's company sold 52 crates of lettuce and 37 crates of cauliflower one day, which vegetable produced more profit that day? How much more?

6. Each month, Frank gets a $2.50 bonus for each crate over 1,000 crates he delivered. If Frank delivers 1,020.5 crates this month, will his bonus be more or less than $75? How much more or less?

7. For the first 3 days of the week, Frank's truck used 5.7 gallons of gas per day. The rest of the week, the truck used 6.8 gallons per day. He drove a total of 521 miles during the week. What was the gas mileage for the week?

8. In one week, Frank drove 638 miles while making deliveries. The following week, he drove 199 fewer miles. The truck usually gets 18 miles per gallon. How many gallons of gas did he use the second week?

Use with pages 196–197.

PRACTICE Fractions and Equivalent Fractions

Write a fraction for the shaded part of each shape or set of shapes.

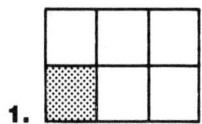 1. _____ 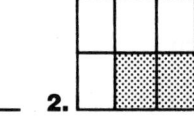 2. _____  3. _____ 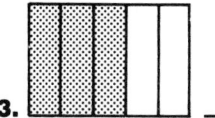 4. _____

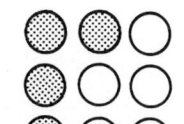 5. _____ 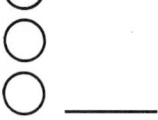 6. _____  7. _____  8. _____

 9. _____ 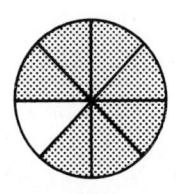 10. _____ 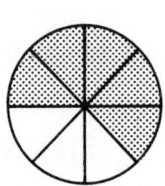 11. _____ 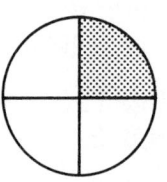 12. _____

Complete.

13. $\frac{1}{3} = \frac{}{15}$
14. $\frac{4}{5} = \frac{16}{}$
15. $\frac{5}{6} = \frac{15}{}$
16. $\frac{2}{7} = \frac{}{56}$

17. $\frac{3}{7} = \frac{}{21}$
18. $\frac{4}{5} = \frac{8}{}$
19. $\frac{7}{8} = \frac{49}{}$
20. $\frac{11}{12} = \frac{}{84}$

21. $\frac{9}{10} = \frac{}{80}$
22. $\frac{2}{3} = \frac{}{21}$
23. $\frac{15}{30} = \frac{60}{}$
24. $\frac{5}{7} = \frac{25}{}$

25. $\frac{16}{32} = \frac{32}{}$
26. $\frac{2}{6} = \frac{}{12}$
27. $\frac{7}{15} = \frac{35}{}$
28. $\frac{6}{21} = \frac{}{7}$

29. $\frac{13}{17} = \frac{39}{}$
30. $\frac{1}{9} = \frac{10}{}$
31. $\frac{24}{48} = \frac{}{8}$
32. $\frac{33}{48} = \frac{11}{}$

Compare. Write = or ≠ for ◯.

33. $\frac{2}{3}$ ◯ $\frac{12}{18}$
34. $\frac{1}{6}$ ◯ $\frac{3}{12}$
35. $\frac{4}{5}$ ◯ $\frac{23}{30}$
36. $\frac{2}{7}$ ◯ $\frac{10}{35}$

37. $\frac{5}{6}$ ◯ $\frac{15}{18}$
38. $\frac{3}{11}$ ◯ $\frac{15}{33}$
39. $\frac{5}{9}$ ◯ $\frac{25}{45}$
40. $\frac{6}{13}$ ◯ $\frac{18}{39}$

PRACTICE Simplifying Fractions

Complete.

1. $\frac{4}{6} = \frac{\Box}{3}$
2. $\frac{3}{9} = \frac{1}{\Box}$
3. $\frac{12}{16} = \frac{\Box}{4}$
4. $\frac{20}{30} = \frac{\Box}{3}$
5. $\frac{9}{18} = \frac{1}{\Box}$
6. $\frac{36}{42} = \frac{\Box}{7}$
7. $\frac{45}{60} = \frac{3}{\Box}$
8. $\frac{27}{33} = \frac{9}{\Box}$
9. $\frac{45}{95} = \frac{9}{\Box}$
10. $\frac{12}{20} = \frac{\Box}{5}$
11. $\frac{24}{40} = \frac{\Box}{5}$
12. $\frac{7}{49} = \frac{1}{\Box}$
13. $\frac{17}{34} = \frac{\Box}{2}$
14. $\frac{28}{63} = \frac{4}{\Box}$

Write the fraction in simplest form.

15. $\frac{4}{8} =$ _____
16. $\frac{12}{14} =$ _____
17. $\frac{14}{16} =$ _____
18. $\frac{10}{15} =$ _____
19. $\frac{2}{18} =$ _____
20. $\frac{40}{56} =$ _____
21. $\frac{81}{90} =$ _____
22. $\frac{27}{51} =$ _____
23. $\frac{10}{12} =$ _____
24. $\frac{4}{12} =$ _____
25. $\frac{4}{8} =$ _____
26. $\frac{6}{42} =$ _____
27. $\frac{22}{26} =$ _____
28. $\frac{40}{80} =$ _____
29. $\frac{30}{39} =$ _____
30. $\frac{35}{95} =$ _____
31. $\frac{19}{57} =$ _____
32. $\frac{24}{32} =$ _____
33. $\frac{12}{36} =$ _____
34. $\frac{62}{72} =$ _____
35. $\frac{2}{86} =$ _____
36. $\frac{34}{66} =$ _____
37. $\frac{8}{56} =$ _____
38. $\frac{25}{35} =$ _____

Use with pages 200–201.

PRACTICE: Writing Decimals for Fractions

Divide and write a decimal for each. Round your answer to the nearest hundredth.

1. $\frac{3}{8}$ = _____
2. $\frac{3}{40}$ = _____
3. $\frac{3}{4}$ = _____
4. $\frac{5}{16}$ = _____

5. $\frac{19}{20}$ = _____
6. $\frac{7}{8}$ = _____
7. $\frac{11}{20}$ = _____
8. $\frac{12}{25}$ = _____

9. $\frac{27}{35}$ = _____
10. $\frac{13}{80}$ = _____
11. $\frac{5}{7}$ = _____
12. $\frac{9}{18}$ = _____

13. $\frac{8}{36}$ = _____
14. $\frac{15}{75}$ = _____
15. $\frac{30}{90}$ = _____
16. $\frac{9}{80}$ = _____

17. $\frac{30}{33}$ = _____
18. $\frac{25}{52}$ = _____
19. $\frac{65}{81}$ = _____
20. $\frac{8}{41}$ = _____

21. $\frac{15}{16}$ = _____
22. $\frac{5}{8}$ = _____
23. $\frac{11}{16}$ = _____
24. $\frac{12}{62}$ = _____

25. $\frac{15}{25}$ = _____
26. $\frac{18}{20}$ = _____
27. $\frac{2}{15}$ = _____
28. $\frac{4}{92}$ = _____

29. $\frac{13}{14}$ = _____
30. $\frac{39}{40}$ = _____
31. $\frac{10}{36}$ = _____
32. $\frac{12}{23}$ = _____

33. $\frac{87}{93}$ = _____
34. $\frac{6}{8}$ = _____
35. $\frac{20}{50}$ = _____
36. $\frac{29}{78}$ = _____

Solve.

37. For a science project about oceans, several students were asked to prepare saltwater solutions containing 0.5625 salt. Write each of the following fractions as a decimal. Write the name of the student who prepared the correct solution.

Paul $\frac{1}{20}$ = _____   Grace $\frac{9}{16}$ = _____

Joyce $\frac{9}{12}$ = _____   Celina $\frac{7}{8}$ = _____

Kevin $\frac{3}{125}$ = _____   Kenny $\frac{2}{16}$ = _____

PRACTICE Mixed Numbers and Fractions

Write a fraction and either a mixed number or a whole number to describe the shaded regions.

1. 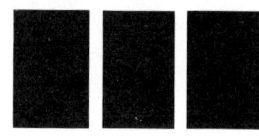 _____

2. 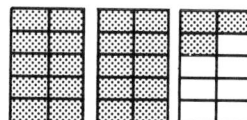 _____

3. 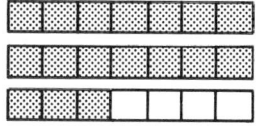 _____

4. 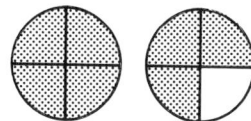 _____

Write each fraction as a whole number or as a mixed number.

5. $\frac{7}{6}$ = _____
6. $\frac{11}{4}$ = _____
7. $\frac{17}{10}$ = _____
8. $\frac{52}{26}$ = _____

9. $\frac{76}{13}$ = _____
10. $\frac{25}{4}$ = _____
11. $\frac{30}{15}$ = _____
12. $\frac{17}{9}$ = _____

13. $\frac{11}{8}$ = _____
14. $\frac{43}{15}$ = _____
15. $\frac{99}{20}$ = _____
16. $\frac{89}{27}$ = _____

17. $\frac{77}{4}$ = _____
18. $\frac{33}{8}$ = _____
19. $\frac{84}{4}$ = _____
20. $\frac{28}{21}$ = _____

21. $\frac{75}{15}$ = _____
22. $\frac{26}{7}$ = _____
23. $\frac{85}{6}$ = _____
24. $\frac{48}{41}$ = _____

Write each as a fraction.

25. $3\frac{8}{5}$ = _____
26. $1\frac{7}{12}$ = _____
27. $14\frac{1}{2}$ = _____
28. $2\frac{7}{8}$ = _____

29. $7\frac{3}{4}$ = _____
30. 25 = _____
31. $5\frac{6}{7}$ = _____
32. $3\frac{9}{10}$ = _____

33. $4\frac{3}{8}$ = _____
34. $9\frac{2}{5}$ = _____
35. $15\frac{1}{3}$ = _____
36. $8\frac{1}{6}$ = _____

37. $2\frac{3}{4}$ = _____
38. $3\frac{5}{7}$ = _____
39. 13 = _____
40. $3\frac{2}{9}$ = _____

Use with pages 204–205.

PRACTICE: Comparing and Ordering Fractions

Compare. Write >, <, or = for ◯.

1. $\frac{1}{2}$ ◯ $\frac{1}{4}$
2. $\frac{3}{4}$ ◯ $\frac{4}{5}$
3. $\frac{4}{6}$ ◯ $\frac{3}{4}$
4. $1\frac{1}{3}$ ◯ $1\frac{6}{9}$
5. $1\frac{1}{4}$ ◯ $1\frac{1}{3}$
6. $\frac{3}{7}$ ◯ $\frac{4}{6}$
7. $\frac{5}{6}$ ◯ $1\frac{1}{6}$
8. $\frac{3}{4}$ ◯ $\frac{5}{6}$
9. $\frac{7}{12}$ ◯ $\frac{2}{3}$
10. $1\frac{7}{12}$ ◯ $\frac{12}{15}$
11. $\frac{10}{12}$ ◯ $\frac{3}{4}$
12. $1\frac{6}{7}$ ◯ $1\frac{3}{4}$
13. $\frac{4}{100}$ ◯ $\frac{2}{50}$
14. $1\frac{4}{5}$ ◯ $1\frac{5}{6}$
15. $3\frac{3}{4}$ ◯ $3\frac{5}{6}$
16. $\frac{7}{12}$ ◯ $1\frac{5}{9}$
17. $\frac{7}{9}$ ◯ $\frac{9}{11}$
18. $\frac{5}{7}$ ◯ $\frac{10}{14}$
19. $\frac{3}{7}$ ◯ $\frac{10}{21}$
20. $\frac{9}{15}$ ◯ $\frac{4}{7}$
21. $2\frac{1}{8}$ ◯ $1\frac{1}{9}$
22. $\frac{8}{15}$ ◯ $\frac{7}{14}$
23. $\frac{4}{7}$ ◯ $\frac{5}{9}$
24. $\frac{1}{11}$ ◯ $\frac{2}{12}$

Order from the greatest to the least.

25. $\frac{5}{8}, \frac{3}{4}, \frac{1}{3}$ _____

26. $7\frac{2}{5}, \frac{37}{7}, 6$ _____

27. $\frac{37}{6}, 6\frac{1}{2}, \frac{19}{3}$ _____

28. $\frac{4}{9}, \frac{2}{3}, \frac{5}{27}$ _____

Order from the least to the greatest.

29. $\frac{56}{8}, 6\frac{13}{16}, 7\frac{1}{4}$ _____

30. $\frac{6}{7}, \frac{11}{14}, \frac{25}{28}$ _____

31. $5\frac{12}{13}, 5\frac{21}{26}, \frac{75}{13}$ _____

32. $\frac{46}{9}, 5\frac{1}{3}, \frac{33}{6}$ _____

Solve.

33. Gladys uses $\frac{2}{3}$ cup of nuts in her cake recipe. Ben uses $\frac{3}{4}$ cup of nuts in his recipe. Who uses more nuts Gladys or Ben? _____

PRACTICE   Estimation

Choose two fractions that are

$\dfrac{11}{20}; \dfrac{1}{12}; \dfrac{32}{33}; \dfrac{18}{19}; \dfrac{7}{12}; \dfrac{2}{33}$

1. close to 0. _____

2. close to $\dfrac{1}{2}$. _____

3. close to 1. _____

Estimate. Write > or < for ◯.

4. $\dfrac{14}{15} + \dfrac{11}{18}$ ◯ 2

5. $2\dfrac{1}{16} + 3\dfrac{3}{7}$ ◯ 5

6. $\dfrac{1}{10} + \dfrac{1}{20}$ ◯ 1

7. $3\dfrac{4}{11} + 6\dfrac{3}{16}$ ◯ 10

8. $\dfrac{11}{12} + \dfrac{1}{2}$ ◯ 1

9. $\dfrac{17}{18} + \dfrac{29}{30} + \dfrac{8}{9}$ ◯ 3

10. $\dfrac{5}{8} + \dfrac{1}{2} + 4\dfrac{1}{6}$ ◯ 5

11. $6\dfrac{7}{8} + 6\dfrac{1}{9}$ ◯ 12

12. $4\dfrac{1}{10} + 5\dfrac{1}{9} + 8\dfrac{7}{8}$ ◯ 20

Estimate.

13. $\dfrac{1}{15} + \dfrac{9}{20}$ about _____

14. $3\dfrac{3}{7} + 2\dfrac{1}{20}$ about _____

15. $\dfrac{7}{13} + \dfrac{7}{15} + 1\dfrac{19}{20}$ about _____

16. $6\dfrac{2}{11} + 6\dfrac{14}{15} + 6\dfrac{1}{2}$ about _____

17. $\dfrac{13}{30} + \dfrac{23}{40}$ about _____

18. $3\dfrac{3}{5} + \dfrac{9}{11} + 2\dfrac{8}{9}$ about _____

19. $\dfrac{1}{16} + \dfrac{1}{17} + \dfrac{8}{19}$ about _____

20. $\dfrac{2}{39} + \dfrac{19}{40}$ about _____

Solve.

21. Addie the Adder likes to have fraction sums in order from the least to the greatest. One sum is out of order. Which sum is it? _____

$\dfrac{1}{15} + \dfrac{1}{14}$; $\dfrac{5}{8} + \dfrac{2}{27}$; $1\dfrac{2}{21} + \dfrac{1}{60}$; $1\dfrac{4}{5} + \dfrac{7}{8}$; $\dfrac{14}{15} + \dfrac{7}{16}$; $1\dfrac{5}{6} + \dfrac{10}{11}$; $2\dfrac{1}{10} + 1\dfrac{7}{8}$

Use with pages 208–209.

PRACTICE Adding Fractions

Add. Write each answer in simplest form.

1. $\dfrac{2}{9} + \dfrac{2}{3} =$ _____

2. $\dfrac{1}{6} + \dfrac{3}{7} =$ _____

3. $\dfrac{4}{5} + \dfrac{1}{8} =$ _____

4. $\dfrac{1}{12} + \dfrac{5}{8} =$ _____

5. $\dfrac{1}{3} + \dfrac{4}{10} =$ _____

6. $\dfrac{1}{4} + \dfrac{1}{5} =$ _____

7. $\dfrac{1}{3} + \dfrac{2}{8} + \dfrac{1}{6} =$ _____

8. $\dfrac{2}{7} + \dfrac{2}{9} + \dfrac{1}{3} =$ _____

9. $\dfrac{3}{10} + \dfrac{2}{5} + \dfrac{2}{8} =$ _____

10. $\dfrac{1}{11} + \dfrac{2}{11} + \dfrac{2}{11} =$ _____

11. $\dfrac{1}{15} + \dfrac{4}{6} =$ _____

12. $\dfrac{1}{3} + \dfrac{3}{8} + \dfrac{2}{12} =$ _____

13. $\dfrac{2}{9} + \dfrac{4}{7}$

14. $\dfrac{3}{8} + \dfrac{4}{10}$

15. $\dfrac{7}{18} + \dfrac{1}{3} + \dfrac{2}{8}$

16. $\dfrac{4}{9} + \dfrac{5}{12}$

17. $\dfrac{2}{7} + \dfrac{1}{8} + \dfrac{1}{2}$

18. $\dfrac{2}{3} + \dfrac{2}{23}$

19. $\dfrac{2}{3} + \dfrac{1}{6}$

20. $\dfrac{3}{8} + \dfrac{5}{9}$

21. $\dfrac{3}{8} + \dfrac{1}{4} + \dfrac{2}{6}$

22. $\dfrac{3}{5} + \dfrac{1}{9} + \dfrac{2}{15}$

23. $\dfrac{1}{6} + \dfrac{1}{2} + \dfrac{1}{4}$

24. $\dfrac{4}{9} + \dfrac{5}{11}$

25. $\dfrac{3}{15} + \dfrac{2}{15} + \dfrac{1}{15}$

26. $\dfrac{6}{7} + \dfrac{1}{8}$

27. $\dfrac{2}{9} + \dfrac{2}{10} + \dfrac{1}{3}$

28. $\dfrac{2}{8} + \dfrac{3}{7}$

29. $\dfrac{3}{12} + \dfrac{4}{6}$

30. $\dfrac{1}{3} + \dfrac{2}{7} + \dfrac{4}{14}$

31. $\dfrac{4}{9} + \dfrac{2}{8} + \dfrac{1}{4}$

32. $\dfrac{1}{9} + \dfrac{3}{24}$

PRACTICE Subtracting Fractions

Subtract. Write the difference in simplest form.

1. $\dfrac{7}{9} - \dfrac{3}{9}$
2. $\dfrac{11}{12} - \dfrac{3}{4}$
3. $\dfrac{26}{38} - \dfrac{10}{38}$
4. $\dfrac{4}{9} - \dfrac{2}{9}$
5. $\dfrac{6}{7} - \dfrac{5}{7}$

6. $\dfrac{8}{8} - \dfrac{1}{4}$
7. $\dfrac{3}{4} - \dfrac{10}{16}$
8. $\dfrac{6}{8} - \dfrac{3}{6}$
9. $\dfrac{2}{3} - \dfrac{1}{3}$
10. $\dfrac{5}{8} - \dfrac{3}{8}$

11. $\dfrac{9}{21} - \dfrac{2}{7}$
12. $\dfrac{2}{5} - \dfrac{1}{3}$
13. $\dfrac{3}{6} - \dfrac{5}{18}$
14. $\dfrac{9}{15} - \dfrac{7}{30}$
15. $\dfrac{5}{12} - \dfrac{3}{9}$

16. $\dfrac{5}{14} - \dfrac{2}{14} =$ _____
17. $\dfrac{2}{81} - \dfrac{2}{81} =$ _____
18. $\dfrac{21}{42} - \dfrac{3}{7} =$ _____

19. $\dfrac{11}{36} - \dfrac{3}{12} =$ _____
20. $\dfrac{6}{20} - \dfrac{4}{15} =$ _____
21. $\dfrac{13}{42} - \dfrac{2}{42} =$ _____

22. $\dfrac{13}{18} - \dfrac{9}{18} =$ _____
23. $\dfrac{20}{22} - \dfrac{3}{11} =$ _____
24. $\dfrac{5}{9} - \dfrac{45}{81} =$ _____

25. $\dfrac{5}{35} - \dfrac{3}{25} =$ _____
26. $\dfrac{12}{76} - \dfrac{1}{76} =$ _____
27. $\dfrac{21}{80} - \dfrac{5}{80} =$ _____

Solve.

28. Alma runs $1\dfrac{1}{2}$ miles every morning with her dog, Mystery. Last Thursday, Alma stopped to talk to Nancy after running only $\dfrac{5}{8}$ mile. How many miles did Alma and Mystery have left to run?

_____

Use with pages 212–213.

PRACTICE — Adding Mixed Numbers

Add. Write the sum in simplest form.

1. $1\frac{2}{6}$
   $+\ 5\frac{3}{6}$

2. $1\frac{2}{12}$
   $+\ 7\frac{9}{12}$

3. $4\frac{4}{9}$
   $+\ 3\frac{4}{9}$

4. $9\frac{5}{12}$
   $+\ 9\frac{1}{6}$

5. $7\frac{3}{12}$
   $+\ 5\frac{6}{12}$

6. $2\frac{1}{4}$
   $+\ 1$

7. $8\frac{1}{8}$
   $+\ 6\frac{3}{8}$

8. $9\frac{2}{7}$
   $+\ 8\frac{4}{7}$

9. $7\frac{3}{9}$
   $+\ 4\frac{2}{9}$

10. $6\frac{1}{10}$
    $+\ 6\frac{2}{10}$

11. $2\frac{1}{6}$
    $7\frac{1}{12}$
    $+\ 4\frac{1}{3}$

12. $8\frac{1}{12}$
    $6\frac{5}{12}$
    $+\ 1\frac{2}{12}$

13. $2\frac{1}{5}$
    $9\frac{1}{5}$
    $+\ 6\frac{3}{10}$

14. $3\frac{1}{5}$
    $8\frac{3}{10}$
    $+\ 3\frac{3}{20}$

15. $6\frac{1}{4}$
    $5\frac{2}{20}$
    $+\ 2\frac{3}{10}$

16. $4\frac{2}{7}$
    $9\frac{2}{7}$
    $1\frac{2}{7}$

17. $4\frac{2}{7}$
    $5\frac{3}{7}$
    $+\ 7\frac{3}{21}$

18. $5\frac{1}{8}$
    $2\frac{5}{16}$
    $+\ 7\frac{1}{4}$

19. $3\frac{4}{8}$
    $3\frac{1}{8}$
    $+\ 1\frac{1}{4}$

20. $1\frac{5}{12}$
    $4\frac{1}{12}$
    $+\ 5\frac{1}{12}$

21. $3\frac{8}{16} + 8\frac{5}{16} =$ _____

22. $4\frac{14}{20} + 1\frac{1}{20} =$ _____

23. $5\frac{2}{8} + 7\frac{2}{8} =$ _____

24. $4\frac{3}{8} + 1 =$ _____

25. $3\frac{4}{8} + 8\frac{1}{8} =$ _____

26. $6\frac{4}{6} + 9\frac{1}{12} =$ _____

27. $1\frac{2}{4} + 7\frac{7}{20} =$ _____

28. $6\frac{2}{9} + 6\frac{3}{9} =$ _____

29. $2\frac{1}{14} + 1\frac{5}{7} =$ _____

PRACTICE: Subtracting Mixed Numbers

Subtract. Write the answer in simplest form.

1. $10\frac{3}{4} - 3\frac{3}{8}$

2. $6\frac{2}{3} - 2\frac{1}{6}$

3. $10\frac{5}{12} - 5\frac{2}{6}$

4. $6\frac{3}{5} - 2\frac{1}{4}$

5. $4\frac{6}{5} - 2\frac{2}{3}$

6. $4\frac{2}{3} - 1\frac{4}{9}$

7. $9\frac{15}{18} - 8\frac{1}{6}$

8. $8\frac{5}{16} - 4\frac{1}{4}$

9. $10\frac{7}{9} - 2\frac{1}{18}$

10. $6\frac{11}{12} - 5\frac{1}{3}$

11. $6\frac{11}{16} - 1\frac{5}{8}$

12. $9\frac{5}{7} - 2\frac{1}{2}$

13. $8\frac{6}{10} - 7\frac{1}{5}$

14. $6\frac{7}{8} - 2\frac{3}{4}$

15. $8\frac{3}{5} - 3\frac{1}{3}$

16. $6\frac{9}{14} - 2\frac{1}{7}$

17. $7\frac{3}{4} - 4\frac{3}{8}$

18. $9\frac{10}{12} - 3\frac{5}{12}$

19. $6\frac{6}{7} - 5\frac{2}{9}$

20. $9\frac{7}{9} - 1\frac{1}{6}$

21. $6\frac{2}{3} - 5\frac{7}{15}$

22. $8\frac{9}{12} - 7\frac{7}{10}$

23. $2\frac{9}{10} - 1\frac{3}{20}$

24. $9\frac{5}{12} - 5\frac{1}{4}$

Use with pages 216–217.

PRACTICE: Adding and Subtracting with Renaming

Add or subtract. Write the answer in simplest form.

1. $7 + 2\frac{3}{8} =$ _____
2. $4\frac{1}{4} + 3\frac{1}{4} =$ _____
3. $\frac{5}{8} + 1\frac{3}{8} =$ _____
4. $5\frac{1}{2} - 2\frac{1}{4} =$ _____
5. $6 - 3\frac{1}{6} =$ _____
6. $8\frac{5}{16} - 4\frac{7}{16} =$ _____
7. $4\frac{1}{10} + 8\frac{2}{5} =$ _____
8. $9\frac{1}{3} + 3\frac{1}{2} =$ _____
9. $7\frac{1}{8} - 6\frac{1}{4} =$ _____
10. $3\frac{2}{5} - 2\frac{1}{4} =$ _____
11. $12\frac{1}{2} + 4\frac{3}{10} =$ _____
12. $10\frac{1}{3} - 7\frac{2}{3} =$ _____

13. $16 - \frac{8}{13}$
14. $8\frac{5}{6} - 6\frac{1}{3}$
15. $11\frac{1}{2} - 3\frac{3}{7}$
16. $5\frac{9}{10} + 6\frac{1}{3}$
17. $4\frac{1}{2} - 1\frac{5}{8}$
18. $9\frac{1}{4} + 5\frac{1}{3}$

19. $24\frac{4}{9} + 8\frac{2}{3}$
20. $7 - 5\frac{5}{8}$
21. $13\frac{3}{7} + 6\frac{4}{7}$
22. $8\frac{1}{2} - 7\frac{3}{4}$
23. $9\frac{2}{5} + 7\frac{3}{10}$
24. $4\frac{7}{9} - 2\frac{1}{9}$

25. $14\frac{7}{8} + 8\frac{1}{8}$
26. $13\frac{1}{3} - 6\frac{5}{6}$
27. $13 + 8\frac{6}{15}$
28. $17\frac{3}{5} - 9\frac{7}{10}$
29. $5\frac{2}{5} + 4\frac{6}{10}$
30. $6\frac{5}{15} - 5\frac{1}{3}$

Solve.

31. Paco was studying the lizards that live in the desert near his house. One lizard measured $7\frac{1}{4}$ in. long. Paco tried to pick up the lizard by the tail. The lizard got away because its tail came off. The tail measured $1\frac{7}{8}$ in. How long was the lizard now? _____

PRACTICE: Solving Multi-step Problems

Write the steps to complete each plan.

1. It takes a shoemaker $1\frac{1}{2}$ hours to repair the soles and $\frac{1}{4}$ hour to repair the heels of a pair of shoes. How many hours will it take the shoemaker to repair 6 soles and 6 heels?

   Step 1: Find the number of hours to repair the soles.

   Step 2: _____

   Step 3: _____

2. Harold can polish and shine 3 pairs of shoes in $\frac{1}{2}$ hour. He receives $1.25 for each pair of shined shoes. How much does Harold receive if he works at this rate for 4 hours?

   Step 1: _____

   Step 2: _____

   Step 3: Multiply $1.25 by the total number of pairs of shoes.

Solve. Make a plan if you need to.

3. Debbie works in a shoe store. On weekdays, she is paid $5.00 per hour for 35 hours. If she works on Saturday, she receives $6.50 per hour. How much does Debbie receive if she works every weekday and 4 hours on Saturday?

4. Debbie sells approximately 50 pairs of socks each week. The store pays $1.25 for each pair of socks and sells them for $2.00 a pair. How much profit does the store make from these sales in 2 weeks?

5. The shoe store bought 25 pairs of shoes for $30.50 each. Debbie marked each pair at $61.00. She sold 10 pairs at this price. She marked down the remaining shoes and sold them for $49.95 a pair. How much profit did the store make on the shoes?

6. Delmont works in a shoe-repair store. He earns $45 in one day. The store's other expenses are $2\frac{1}{4}$ times Delmont's earnings. How much money must the store receive from sales to break even in a 5-day week?

7. The shoemaker uses $2\frac{3}{4}$ feet of leather strap to make one sandal. There are 15 feet of strap in stock. How many more feet of strap does the shoemaker need to make 4 pairs of sandals?

8. Debbie has to commute $6\frac{1}{2}$ km to work. The bus she takes travels about $3\frac{1}{4}$ km in 10 minutes. About how much time does she spend on the bus going to and from work?

Use with pages 220–221.

PRACTICE: Multiplying Fractions by Fractions

Multiply. Write the answer in simplest form.

1. $\frac{1}{8} \times \frac{4}{5} =$ _____
2. $\frac{3}{5} \times \frac{5}{6} =$ _____
3. $\frac{2}{3} \times \frac{1}{3} =$ _____

4. $\frac{3}{4} \times \frac{1}{2} =$ _____
5. $\frac{2}{7} \times \frac{1}{4} =$ _____
6. $\frac{13}{16} \times \frac{4}{5} =$ _____

7. $\frac{5}{6} \times \frac{2}{3} =$ _____
8. $\frac{9}{10} \times \frac{5}{8} =$ _____
9. $\frac{6}{13} \times \frac{7}{10} =$ _____

10. $\frac{1}{2} \times \frac{13}{16} =$ _____
11. $\frac{5}{9} \times \frac{3}{5} =$ _____
12. $\frac{3}{4} \times \frac{5}{6} =$ _____

13. $\frac{4}{5} \times \frac{3}{4} \times \frac{1}{6} =$ _____
14. $\frac{2}{3} \times \frac{1}{5} \times \frac{5}{7} =$ _____

15. $\frac{9}{10} \times \frac{7}{8} \times \frac{5}{9} =$ _____
16. $\frac{1}{10} \times \frac{1}{5} \times \frac{1}{2} =$ _____

17. $\frac{5}{6} \times \frac{3}{10} \times \frac{6}{13} =$ _____
18. $\frac{7}{9} \times \frac{3}{4} \times \frac{4}{5} =$ _____

Solve.

19. Oliver cuts $\frac{3}{4}$ yard of leather to make a belt. He covers $\frac{5}{9}$ of the leather with a design. What fraction of a yard of leather is covered by the design? _____

20. The buckle that Oliver puts on the belt weighs $\frac{1}{2}$ pound. Silver makes up $\frac{2}{3}$ of that weight. How much does the silver in the buckle weigh? _____

PRACTICE: Multiplying Fractions and Whole Numbers

Multiply. Write the answer in simplest form.

1. $6 \times \frac{1}{4} =$ _____
2. $9 \times \frac{1}{3} =$ _____
3. $10 \times \frac{3}{5} =$ _____
4. $\frac{7}{8} \times 24 =$ _____
5. $\frac{5}{9} \times 15 =$ _____
6. $\frac{3}{4} \times 30 =$ _____
7. $16 \times \frac{4}{5} =$ _____
8. $20 \times \frac{1}{8} =$ _____
9. $\frac{2}{3} \times 28 =$ _____
10. $\frac{2}{7} \times 70 =$ _____
11. $31 \times \frac{1}{9} =$ _____
12. $\frac{5}{6} \times 36 =$ _____
13. $\frac{4}{9} \times 45 =$ _____
14. $5 \times \frac{1}{3} =$ _____
15. $\frac{5}{8} \times 120 =$ _____
16. $39 \times \frac{2}{3} =$ _____
17. $\frac{3}{4} \times 44 =$ _____
18. $25 \times \frac{3}{7} =$ _____
19. $246 \times \frac{4}{5} =$ _____
20. $532 \times \frac{5}{6} =$ _____
21. $\frac{7}{8} \times 144 =$ _____
22. $\frac{3}{6} \times \frac{2}{3} \times 12 =$ _____
23. $\frac{3}{5} \times 9 \times \frac{10}{27} =$ _____
24. $\frac{1}{2} \times \frac{4}{3} \times 16 =$ _____

Solve.

25. The sports store has 50 tennis rackets. Of these, $\frac{3}{5}$ are metal. How many tennis rackets are metal?

26. The store has in stock 128 pairs of running shoes. It has $\frac{5}{8}$ as many pairs of golf shoes. How many pairs of golf shoes are there in stock?

PRACTICE: Estimating a Fraction of a Number

Estimate.

1. $\frac{3}{4} \times 95 =$ _____
2. $\frac{3}{4} \times 193 =$ _____
3. $\frac{3}{4} \times 823 =$ _____
4. $\frac{3}{5} \times 198 =$ _____
5. $\frac{4}{5} \times 490 =$ _____
6. $\frac{2}{5} \times 208 =$ _____
7. $\frac{2}{3} \times 92 =$ _____
8. $\frac{2}{3} \times 212 =$ _____
9. $\frac{2}{3} \times 619 =$ _____
10. $\frac{3}{4} \times 590 =$ _____
11. $\frac{1}{2} \times 717 =$ _____
12. $\frac{4}{5} \times 585 =$ _____
13. $\frac{2}{3} \times 592 =$ _____
14. $\frac{4}{5} \times 802 =$ _____
15. $\frac{2}{3} \times 178 =$ _____
16. $\frac{4}{5} \times 426 =$ _____
17. $\frac{3}{4} \times 907 =$ _____
18. $\frac{3}{4} \times 286 =$ _____
19. $\frac{2}{3} \times 121 =$ _____
20. $\frac{4}{5} \times 198 =$ _____
21. $\frac{3}{5} \times 309 =$ _____
22. $\frac{1}{5} \times 203 =$ _____

Solve.

23. New moccasins cost $23.65. Alex has $\frac{1}{4}$ of the price. Estimate how much money Alex has. _____

24. The 152 members of the Roaring Rodeo Singers are choosing between rhinestones and beads for their costumes. If $\frac{1}{5}$ prefer rhinestone decorations, about how many prefer the beads? _____

PRACTICE: Estimation/Number Sense

Mr. Sardi prepared and planted a large garden this year. He decides to set up a roadside stand and sell his vegetables. He has to decide how much to sell and what to charge for each kind of vegetable.

Decide whether or not each question needs to be answered. If it does, decide if an exact answer or only an estimate can be found. Write *need not answer, can get an exact answer,* or *can get an estimate only.*

1. How much money was needed to prepare the garden and buy the seeds and plants?

2. How much time was needed to prepare, plant, and tend the garden?

3. Where shall Mr. Sardi put the roadside stand?

4. How many boxes and packages for the vegetables should he buy?

5. How much space will Mr. Sardi need for displaying the vegetables?

6. How much should Mr. Sardi charge for each kind of vegetable in order to make $250 profit?

7. How much will it cost to build a roadside stand?

8. How much should Mr. Sardi pay someone to work at the stand?

9. Should Mr. Sardi sell tomatoes by the pound or individually?

10. How much would it cost Mr. Sardi to advertise?

11. How much did Mr. Sardi make the first day?

12. How much money will Mr. Sardi make in one week?

13. Should Mr. Sardi open the stand only during rush hours?

14. How much profit did Mr. Sardi make after one week of sales?

Use with pages 236–237.

PRACTICE: Multiplying Mixed Numbers

Multiply. Write the answer in simplest form.

1. $7\frac{1}{3} \times 5 = $ _____
2. $1\frac{1}{2} \times 8 = $ _____
3. $1\frac{3}{4} \times 2 = $ _____
4. $3\frac{1}{4} \times \frac{4}{5} = $ _____
5. $4\frac{7}{8} \times 1\frac{1}{3} = $ _____
6. $\frac{1}{5} \times 2\frac{1}{2} = $ _____
7. $4 \times 1\frac{2}{3} = $ _____
8. $3\frac{1}{4} \times 1\frac{1}{3} = $ _____
9. $1\frac{3}{8} \times 7 = $ _____
10. $\frac{5}{6} \times 2\frac{2}{5} = $ _____
11. $2\frac{2}{5} \times 2 = $ _____
12. $2\frac{5}{6} \times 3\frac{3}{4} = $ _____
13. $10 \times 1\frac{7}{8} = $ _____
14. $2\frac{1}{4} \times 1\frac{2}{5} = $ _____
15. $2\frac{1}{8} \times 4\frac{4}{5} = $ _____
16. $3\frac{2}{3} \times 4\frac{1}{2} = $ _____
17. $5\frac{1}{2} \times \frac{1}{4} = $ _____
18. $4\frac{1}{5} \times 1\frac{1}{2} = $ _____
19. $8\frac{1}{2} \times 8\frac{1}{2} = $ _____
20. $14\frac{4}{5} \times \frac{2}{3} = $ _____
21. $11 \times 3\frac{2}{7} = $ _____
22. $4\frac{2}{3} \times 1\frac{1}{6} \times \frac{1}{7} = $ _____
23. $5 \times 1\frac{3}{8} \times \frac{3}{5} = $ _____
24. $16 \times 4\frac{1}{2} \times 1\frac{1}{3} = $ _____

Solve.

25. Karen ran 1 mile in $5\frac{1}{2}$ minutes. It took Cindy $1\frac{1}{4}$ times as long to run the mile. What was Cindy's time?

_____

26. Ricardo jogged $4\frac{3}{5}$ miles. Hector jogged $1\frac{2}{3}$ times as far. How far did Hector jog?

_____

Use with pages 238–239.

PRACTICE: Dividing Whole Numbers and Fractions

Write the reciprocal of each number.

1. $\frac{1}{5}$ _____
2. 18 _____
3. $\frac{2}{9}$ _____
4. $\frac{13}{17}$ _____
5. $1\frac{3}{4}$ _____
6. $11\frac{1}{2}$ _____
7. $5\frac{5}{6}$ _____
8. $11\frac{1}{9}$ _____

Divide. Write the answer in simplest form.

9. $8 \div \frac{8}{3} =$ _____
10. $\frac{3}{4} \div \frac{5}{8} =$ _____
11. $\frac{1}{2} \div 10 =$ _____
12. $30 \div \frac{3}{2} =$ _____
13. $\frac{4}{5} \div 12 =$ _____
14. $\frac{3}{4} \div \frac{1}{6} =$ _____
15. $\frac{3}{7} \div 6 =$ _____
16. $19 \div \frac{1}{2} =$ _____
17. $\frac{4}{9} \div 8 =$ _____
18. $49 \div \frac{7}{2} =$ _____
19. $\frac{10}{3} \div \frac{5}{6} =$ _____
20. $\frac{7}{2} \div \frac{3}{8} =$ _____
21. $\frac{2}{7} \div 8 =$ _____
22. $6 \div \frac{3}{4} =$ _____
23. $10 \div \frac{2}{5} =$ _____
24. $\frac{9}{10} \div 7 =$ _____
25. $60 \div \frac{2}{5} =$ _____
26. $\frac{2}{3} \div 6 =$ _____

Solve.

27. Workers are laying pieces of sod around the West End apartment complex. One stack of sod is 24 in. high. Each piece of sod is $\frac{2}{3}$ in. high. How many pieces of sod are there in the stack?

    _____

28. It takes a worker $\frac{3}{4}$ hour to mow one section of lawn by the West End apartment complex. How many sections can the worker mow in 6 hours?

    _____

Use with pages 240–241.

PRACTICE: Dividing Fractions and Mixed Numbers

Divide. Write the answer in simplest form.

1. $8\frac{1}{2} \div 2 =$ _____
2. $15\frac{2}{3} \div 3 =$ _____
3. $21\frac{3}{4} \div 7\frac{1}{4} =$ _____
4. $10\frac{1}{2} \div \frac{1}{4} =$ _____
5. $25\frac{2}{5} \div \frac{2}{5} =$ _____
6. $50\frac{1}{2} \div 5 =$ _____
7. $14\frac{1}{3} \div \frac{2}{3} =$ _____
8. $2\frac{1}{4} \div \frac{3}{4} =$ _____
9. $45\frac{5}{9} \div 9\frac{1}{9} =$ _____
10. $40\frac{3}{5} \div 8\frac{3}{5} =$ _____
11. $10\frac{1}{8} \div 5 =$ _____
12. $20\frac{3}{4} \div \frac{3}{4} =$ _____
13. $\frac{3}{4} \div 3\frac{1}{2} =$ _____
14. $\frac{1}{2} \div 7\frac{1}{2} =$ _____
15. $\frac{2}{3} \div 5\frac{1}{3} =$ _____
16. $\frac{3}{9} \div 2\frac{1}{3} =$ _____
17. $\frac{1}{4} \div 15\frac{1}{4} =$ _____
18. $\frac{5}{6} \div 8\frac{1}{3} =$ _____
19. $\frac{3}{5} \div 6\frac{1}{5} =$ _____
20. $\frac{7}{8} \div 11\frac{1}{4} =$ _____
21. $\frac{1}{2} \div 13\frac{1}{4} =$ _____
22. $\frac{1}{3} \div 4\frac{1}{3} =$ _____
23. $\frac{7}{10} \div 10\frac{1}{5} =$ _____
24. $\frac{3}{4} \div 1\frac{3}{4} =$ _____

Use the clues.
Find the mystery number.

- Divide me by $\frac{1}{4}$ and the answer is 25.
- Divide me by 5 and the answer is $1\frac{1}{4}$.

Who am I?

_____

Use with pages 242–243.

PRACTICE: Interpreting the Quotient and the Remainder

Tell what you would do with the quotient or the remainder in each problem. Write the letter of the correct answer.

1. There are 96 people associated with the West Side Soccer League. If each team has 15 players and the rest are coaches, how many people are coaches?

   a. 6   b. $6\frac{6}{15}$ c. 15

2. Before the soccer season started, all 96 people met at the community center and drove to the different playing fields. If 7 people rode in one car, how many cars were used?

   a. 13   b. $13\frac{5}{7}$ c. 14

Solve.

3. Each player wears a league shirt. There is room for a 2-digit number on the back of each shirt. The league has 197 sew-on numbers. How many shirts with 2-digit numbers can be made?

4. Before distribution to the players and coaches, the shirts were packaged into boxes. If 7 shirts fit in each box, how many boxes will be needed for 96 shirts? _____

5. Each year, the league buys netting to make goals. This year, the league bought 225 feet of netting. Each goal is made of 36 feet of netting. How many feet of netting were left over after the 6 goals were made? _____

6. Some of the goals need new posts. Each post measures 7 feet long. The lumberyard has a sale on posts 15 feet long. How many goal posts can be made from 4 of the posts on sale at the lumberyard? _____

7. The teams keep the soccer balls in net bags. Each bag holds 12 balls. The league has 66 soccer balls available for the season. How many bags are needed to carry all the soccer balls?

8. The league has available 27 referees to officiate at the games. For every 3 referees at a game, there are 2 linesmen. How many linesmen are available? _____

9. Workers are replacing the old seats in a 12-foot section of the grandstand. Each of the new seats is 23 inches wide. How many new seats can they install?

10. Each player on one team is given 3 pairs of game shorts. The team has 40 pairs of shorts. How many players can be outfitted? How many more pairs of shorts are needed to outfit all 15 players? _____

Use with pages 244–245.

PRACTICE Inches

Measure to the nearest $\frac{1}{4}$ inch.

1. _____ 2. _____
   _____ in.   _____ in.

3. _____ 4. _____
   _____ in.   _____ in.

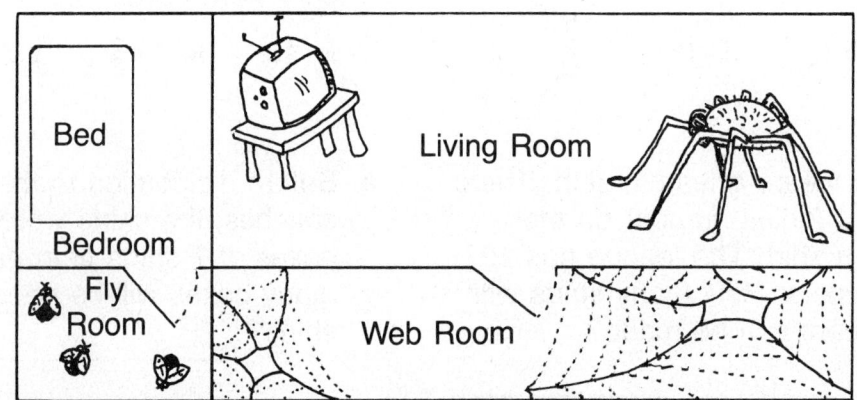

Above is the floor plan for a spider's apartment. Measure the distance to the nearest $\frac{1}{16}$ inch.

5. length of apartment = _____
6. width of apartment = _____
7. length of fly room = _____
8. width of living room = _____
9. length of bed = _____
10. length of web room = _____

Draw a line to the given measure.

11. 2 in.
12. $1\frac{3}{4}$ in.
13. $\frac{1}{3}$ in.
14. $1\frac{1}{2}$ in.
15. 1 in.
16. $\frac{2}{3}$ in.
17. $2\frac{1}{8}$ in.
18. $1\frac{1}{4}$ in.
19. $\frac{1}{2}$ in.
20. $1\frac{1}{8}$ in.
21. $2\frac{1}{3}$ in.
22. $\frac{3}{4}$ in.
23. $1\frac{5}{8}$ in.
24. $1\frac{1}{3}$ in.
25. $\frac{3}{8}$ in.

100   Use with pages 246–247.

PRACTICE Customary Units of Length

Complete.

1. 30 in. = _____ ft
2. 5 ft 3 in. = _____ ft
3. $1\frac{3}{4}$ yd = _____ ft
4. 96 in. = _____ yd
5. $5\frac{2}{3}$ yd = _____ ft
6. 3 ft 2 in. = _____ ft
7. 330 ft = _____ mi
8. 92 ft = _____ yd
9. 7,040 ft = _____ mi
10. 220 yd = _____ mi
11. 1,100 yd = _____ mi
12. 528 ft = _____ mi
13. $2\frac{1}{3}$ ft = _____ yd
14. 6 ft 6 in. = _____ yd
15. $4\frac{4}{9}$ yd = _____ ft
16. 30 in. = _____ yd
17. 3,960 ft = _____ mi
18. 106 in. = _____ ft

Write *true* or *false*.

19. $2\frac{1}{2}$ ft = 2 ft 6 in. _____
20. 5 yd 12 in. = 192 in. _____
21. 44 in. = $1\frac{2}{9}$ yd _____
22. 220 yd = 8 mi _____
23. 2 mi = 3,520 yd _____
24. $3\frac{5}{6}$ ft = 1 yd 30 in. _____

Add or subtract.

25. 3 ft 2 in.
   + 1 ft 10 in.

26. $4\frac{3}{4}$ ft
   + $2\frac{5}{8}$ ft

27. 7 yd 1 ft
   + 3 yd 2 ft

28. 9 yd 1 ft
   +   6 ft

29. $9\frac{3}{4}$ ft
   − 6   ft 2 in.

30. 6 ft 4 in.
   − 4 ft 8 in.

31. 12 yd
   − 7 yd 2 ft

32. 2 mi 220 yd
   + 1 mi 660 yd

33. 3 yd 2 ft
   −   9 in.

Use with pages 248–249.

PRACTICE: Customary Units of Capacity and Weight

Complete.

1. 2 gal = ___ qt
2. 3 qt = ___ pt
3. 8 pt = ___ gal
4. 12 pt = ___ qt
5. 11 pt = ___ c
6. 16 pt = ___ gal
7. 3 gal = ___ c
8. 9 pt = ___ c
9. 3 pt = ___ fl oz
10. 14 pt = ___ qt
11. 4 gal = ___ qt
12. 17 pt = ___ c

Which unit would you use to measure? Write *oz*, *lb*, or *T*.

13.

14.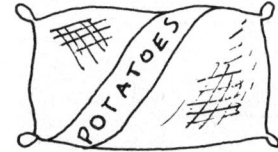

15.

16.

Solve.

17. Tim Smith uses 32 cups plantanos in his recipe for making pies. How many quarts of plantanos does he use?

18. The San San Wong Bakery uses 3,200 oz flour each day to make moon cakes. How many pounds of flour does the San San Wong Bakery use each day?

Use with pages 250–251.

PRACTICE — Units of Time: Addition and Subtraction

Complete.

1. $7\frac{1}{2}$ h = _____ min
2. 240 s = _____ min
3. 45 s = _____ min
4. 400 d = _____ y _____ d
5. 14 wk = _____ d
6. 56 h = _____ d
7. 416 wk = _____ y
8. 3 y 5 mo = _____ mo
9. 87 mo = _____ y _____ mo

10. 6 min 25 s
   + 11 min 18 s

11. 10 d 20 h
   + 8 d 4 h

12. 12 h 15 min
   + 9 h 55 min

13. 8 wk 5 d
   − 4 wk 4 d

14. 2 y 7 mo
   − 1 y 8 mo

15. 14 h 32 min
   −   40 min

16. 5 y
   − 2 y 10 mo

17. 42 h 19 min
   + 35 h 42 min

18. 21 min 27 s
   − 13 min 33 s

19. 23 y 17 wk
   + 59 y 48 wk

20. 9 y 100 d
   − 6 y 270 d

21. 97 d 19 h
   + 45 d 17 h

22. 64 wk 2 d
   − 31 wk 6 d

23. 32 y 11 mo
   + 27 y 1 mo

24. 80 h 9 min
   − 73 h 51 min

Solve.

25. Bryan has worked at Wonderwatches for 6 years 27 weeks. His friend Elsie has worked there 50 fewer weeks. How long has Elsie been at Wonderwatches?

26. Bryan takes 35 minutes for lunch on Monday, Wednesday, and Friday. On the other two days, he takes 1 hour 15 minutes. How much time does Bryan take for lunch in a week?

Use with pages 252–253.

PRACTICE: Temperature

Write the amount of change

1. from 20°F to 45°F. _____
2. from 0°F to 27.6°F. _____
3. from 105°C to 26°C. _____
4. from 41°C to ⁻12°C. _____
5. from 10°F to ⁻15°F. _____
6. from ⁻30°F to ⁻20°F. _____
7. from 68°F to the boiling temperature of water. _____
8. from the freezing temperature of water to 0°F. _____

Write the temperature.

9. 45°F is increased by 32°F. _____
10. 0°F is decreased by 13°F. _____
11. 40°C is decreased by 33°C. _____
12. ⁻9°C is increased by 12°C. _____
13. The boiling temperature of water is increased by 56°F. _____
14. The freezing temperature of water is decreased by 42°F. _____

15.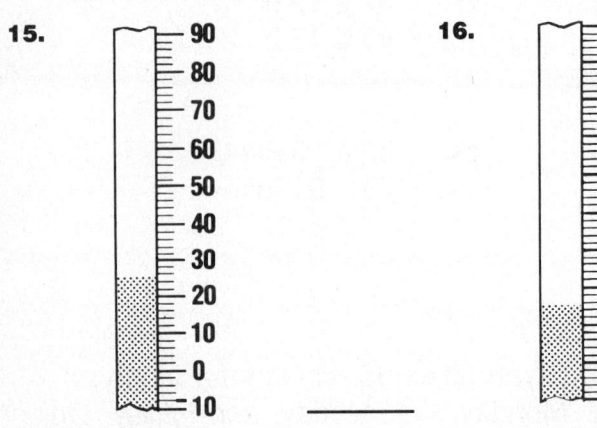
16.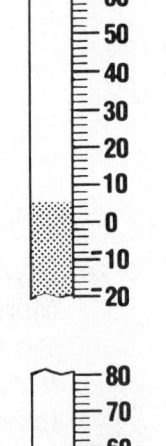
17.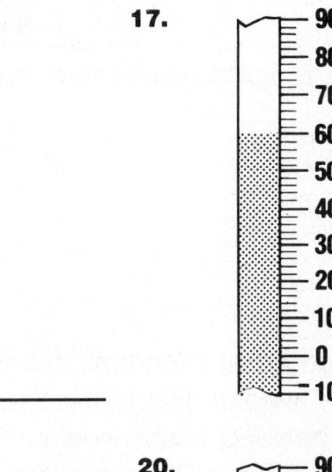

18.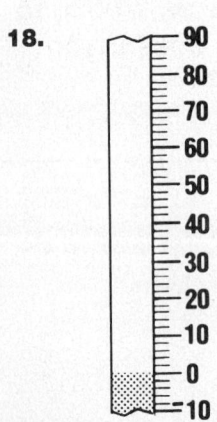
19.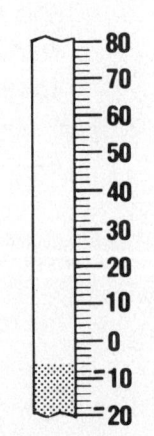
20.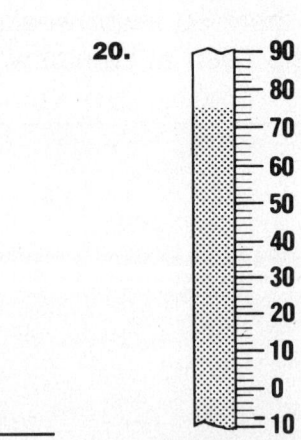

Use with pages 254–255.

PRACTICE Using a Schedule/Time-Zone Map

Use the time-zone map to answer each problem. Write *true* or *false*.

1. Chicago and Denver are in the same time zone. _____

2. It is the same time in Dallas and New Orleans. _____

3. As you travel east, each time zone will be 1 hour earlier than the last one. _____

4. The Mountain time zone and the Central time zone are 1 hour apart. _____

Use the time-zone map and the schedule to solve each problem.

5. Tom lives in Chicago. He has a pen pal in Seattle. When Tom goes to bed at 9:30 P.M., what time is it at his pen pal's house? _____

6. Tom is going to visit his pen pal in Seattle. He leaves Chicago at 9:35 A.M. The flight takes 5 hours and 12 minutes. What time is it in Seattle when Tom gets off the plane? _____

7. It takes 1 hour to get from the Seattle airport to Tom's pen pal's house. If Tom arrives at his pen pal's house at 2:00 P.M., what time is it in Chicago? _____

8. Tom promised to call his parents after he arrived at his pen pal's house. His parents are expecting the phone call at 7:00 P.M. If Tom calls on time, what time will it be in Chicago? _____

9. Tom wants to allow at least 30 minutes to change planes in Denver on the way to Seattle. The flight from Chicago to Denver leaves at 9:35 A.M. and takes 2 hours and 35 minutes. The flight from Denver to Seattle leaves at 11:45 A.M. Does he have enough time to change planes? _____

10. The flight from Seattle to Denver takes 3 hours and 20 minutes. The flight from Denver to Chicago leaves Denver at 1:45 P.M. If Tom wants to allow at least 30 minutes to change planes, what is the latest time his flight can leave Seattle on the way home? _____

Use with pages 256–257.

PRACTICE   Ratios

Write the fraction.

1. 5:8 _____
2. 2:10 _____
3. 7:10 _____
4. 9:2 _____
5. 10:7 _____
6. 4:3 _____
7. 8:3 _____
8. 9:3 _____
9. 10:6 _____
10. 8:6 _____
11. 6:7 _____
12. 10:9 _____
13. 6 to 5 _____
14. 9 to 8 _____
15. 6 to 1 _____
16. 7 to 2 _____
17. 10 to 8 _____
18. 47 to 80 _____
19. 85 to 114 _____
20. 21 to 3 _____
21. 455 to 7 _____
22. 8 quarts for every 2 gallons _____
23. 60 minutes for 1 hour _____
24. 36 months for every 3 years _____
25. 1 week for every 7 days _____

Write the ratio.

26. $\frac{3}{9}$ = _____ to _____
27. $\frac{1}{3}$ = _____ to _____
28. $\frac{1}{5}$ = _____ to _____
29. $\frac{51}{52}$ = _____ to _____
30. $\frac{31}{222}$ = _____ to _____
31. $\frac{18}{9}$ = _____ to _____
32. $\frac{41}{7}$ = _____ to _____
33. $\frac{10}{2}$ = _____ to _____
34. $\frac{17}{412}$ = _____ to _____
35. $\frac{1}{3}$ = _____ : _____
36. $\frac{1}{11}$ = _____ : _____
37. $\frac{13}{50}$ = _____ : _____
38. $\frac{7}{12}$ = _____ : _____
39. $\frac{44}{48}$ = _____ : _____
40. $\frac{15}{30}$ = _____ : _____
41. $\frac{19}{2}$ = _____ : _____
42. $\frac{19}{20}$ = _____ : _____

43. The movie theater sold 95 adult tickets and 61 children's tickets. Write a fraction for the ratio of adult to children's tickets? _____

44. The theater pays $5.00 for 100 paper drinking cups. What rate does the theater pay for cups? _____

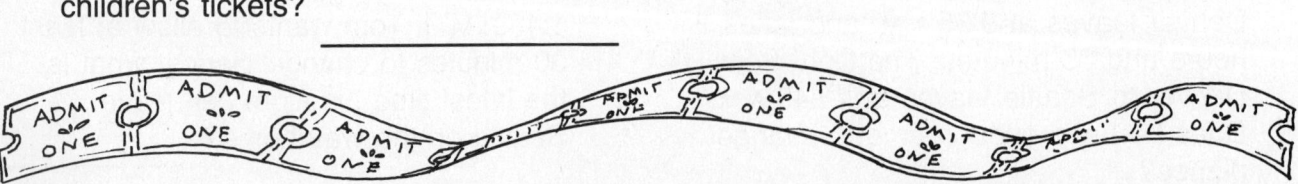

PRACTICE Equal Ratios

Copy and complete the table of equal ratios.

2	4		16	32	64	
5	10	20				320

Solve.

1. $\frac{2}{3} = \frac{6}{\_}$
2. $\frac{3}{8} = \frac{9}{\_}$
3. $\frac{10}{15} = \frac{\_}{3}$
4. $\frac{4}{11} = \frac{12}{\_}$
5. $\frac{4}{7} = \frac{8}{\_}$
6. $\frac{5}{6} = \frac{\_}{36}$
7. $\frac{12}{32} = \frac{\_}{8}$
8. $\frac{14}{18} = \frac{\_}{9}$
9. $\frac{12}{21} = \frac{4}{\_}$
10. $\frac{35}{42} = \frac{\_}{6}$
11. $\frac{7}{8} = \frac{14}{\_}$
12. $\frac{2}{7} = \frac{\_}{21}$
13. $\frac{21}{35} = \frac{\_}{5}$
14. $\frac{4}{9} = \frac{\_}{27}$
15. $\frac{15}{24} = \frac{\_}{8}$
16. $\frac{3}{10} = \frac{15}{\_}$

Use cross products to tell whether the ratios are equal.
Write = or ≠ for ◯.

17. $\frac{12}{9}$ ◯ $\frac{72}{54}$
18. $\frac{14}{8}$ ◯ $\frac{42}{24}$
19. $\frac{20}{8}$ ◯ $\frac{19}{9}$
20. $\frac{2}{3}$ ◯ $\frac{30}{40}$
21. $\frac{18}{10}$ ◯ $\frac{15}{20}$
22. $\frac{19}{17}$ ◯ $\frac{95}{85}$
23. $\frac{9}{6}$ ◯ $\frac{27}{18}$
24. $\frac{1}{4}$ ◯ $\frac{4}{16}$
25. $\frac{20}{6}$ ◯ $\frac{6}{20}$
26. $\frac{15}{6}$ ◯ $\frac{14}{9}$
27. $\frac{17}{21}$ ◯ $\frac{102}{126}$
28. $\frac{2}{11}$ ◯ $\frac{5}{20}$
29. $\frac{6}{7}$ ◯ $\frac{5}{19}$
30. $\frac{13}{8}$ ◯ $\frac{9}{13}$
31. $\frac{19}{12}$ ◯ $\frac{6}{8}$
32. $\frac{3}{13}$ ◯ $\frac{9}{26}$
33. $\frac{2}{12}$ ◯ $\frac{15}{16}$
34. $\frac{18}{14}$ ◯ $\frac{90}{70}$
35. $\frac{4}{20}$ ◯ $\frac{13}{2}$
36. $\frac{17}{6}$ ◯ $\frac{6}{17}$
37. $\frac{15}{8}$ ◯ $\frac{3}{6}$
38. $\frac{1}{7}$ ◯ $\frac{3}{21}$
39. $\frac{3}{4}$ ◯ $\frac{15}{20}$
40. $\frac{7}{15}$ ◯ $\frac{28}{60}$

Use with pages 268–269.

PRACTICE Working Backward

Work backward. Write the letter of the correct answer.

1. Carla, Molly, and Toni received stamps for their stamp collections from their grandmother. Carla gave $\frac{1}{3}$ of these to the Stamp Club. She divided the remaining stamps equally among Molly, Toni, and herself. If Carla's share was 20 stamps, how many did their grandmother send them?

   a. 30   b. 60   c. 90

2. Molly ordered a package of wildlife stamps from a catalog. When Toni saw them, she thought they were so nice she ordered twice as many as Molly. Carla thought they were beautiful and ordered twice as many as Toni. If Carla ordered 24 stamps, how many stamps did Molly receive?

   a. 12   b. 6   c. 48

Work backward to solve.

3. Carla and Toni traded stamps. Carla received 10 stamps from Molly in return for 5 stamps. Molly then traded these stamps for 5 different stamps from Toni. Toni traded $\frac{1}{2}$ the stamps that Carla received from Molly. How many stamps did Toni trade? _____

4. Toni went to a stamp show where she bought many stamps. She gave $\frac{1}{2}$ of these to Molly, who gave $\frac{1}{3}$ of what she received to Carla. Carla received 10 stamps. How many stamps did Toni buy at the stamp show?

   _____

5. Carla bought a rare stamp at the show but discovered Molly already had it. She sold it to another collector and received twice what she paid for it. The collector discovered the actual value of the stamp was twice as much as the price he purchased it for. The stamp was worth $5.00. How much did Carla pay for the stamp?

   _____

Use with pages 270–271.

PRACTICE: Working Backward

Work backward to solve.

6. Carla, Toni, and Molly spend much of their time outdoors. They go camping, canoeing, and hiking with their parents. Last year, $\frac{1}{6}$ of their outings were canoeing trips. They spent twice as many outings camping. The family took the same number of hiking trips as canoeing and camping trips combined. If they went camping 12 times, how many times did they go canoeing and hiking?

7. The cost of renting a canoe for one adult is $7.00. The cost for a person under 16 years old is $\frac{1}{4}$ the amount paid by an adult. If the total paid to rent canoes was $17.50, how many adults and how many people under 16 rented canoes?

8. At a National Park campground, the rangers offered to lead a nature walk. To decide on which day to hold the walk, the rangers surveyed the campers. Half of the people selected Monday. Half of the remaining people selected Tuesday, and the remaining 8 campers selected Wednesday. How many campers were surveyed?

9. Toni has been a camper for three years less than Molly. Carla has camped 4 years longer than Molly. Toni started camping when she was 9 years old. She is now 11 years old. For how many years has Carla been camping? and Molly?

10. Molly, Toni, and Carla went fishing with their father. He caught twice as many fish as Carla caught. Carla caught 4 more fish than Molly caught. Molly caught 8 fewer fish than Toni caught. Toni caught 11 fish. How many fish did each person catch?

Use with pages 270–271.

PRACTICE — Proportions

Solve the proportions.

1. $\frac{n}{12} = \frac{6}{9}$; n = _____
2. $\frac{3}{n} = \frac{4}{12}$; n = _____
3. $\frac{5}{n} = \frac{30}{6}$; n = _____
4. $\frac{6}{7} = \frac{n}{56}$; n = _____
5. $\frac{n}{10} = \frac{14}{35}$; n = _____
6. $\frac{n}{8} = \frac{3}{12}$; n = _____
7. $\frac{6}{n} = \frac{9}{12}$; n = _____
8. $\frac{n}{15} = \frac{6}{10}$; n = _____
9. $\frac{12}{n} = \frac{6}{12}$; n = _____
10. $\frac{n}{20} = \frac{12}{16}$; n = _____
11. $\frac{14}{n} = \frac{42}{12}$; n = _____
12. $\frac{12}{8} = \frac{27}{n}$; n = _____

Write *true* or *false*.

13. $\frac{2}{9} = \frac{14}{64}$ _____
14. $\frac{3}{8} = \frac{9}{24}$ _____
15. $\frac{7}{10} = \frac{8}{25}$ _____
16. $\frac{7}{8} = \frac{28}{32}$ _____
17. $\frac{9}{3} = \frac{3}{1}$ _____
18. $\frac{5}{6} = \frac{3}{4}$ _____
19. $\frac{27}{18} = \frac{3}{2}$ _____
20. $\frac{5}{9} = \frac{8}{16}$ _____
21. $\frac{17}{2} = \frac{35}{4}$ _____
22. $\frac{7}{8} = \frac{21}{24}$ _____
23. $\frac{10}{45} = \frac{8}{30}$ _____
24. $\frac{4}{8} = \frac{5}{10}$ _____
25. $\frac{1}{4} = \frac{1}{3}$ _____
26. $\frac{2}{7} = \frac{5}{28}$ _____
27. $\frac{3}{5} = \frac{4}{6}$ _____
28. $\frac{5}{8} = \frac{25}{40}$ _____
29. $\frac{5}{8} = \frac{25}{49}$ _____
30. $\frac{16}{3} = \frac{10}{2}$ _____

Solve.

31. An actress in the daytime drama *Edge of Evening* spends 25 minutes learning 15 lines of dialogue. At that rate, how long would it take her to learn 120 lines?

32. Electricity on the stage sets costs $540 for each 30 minutes of shooting. How much would electricity cost for 135 minutes?

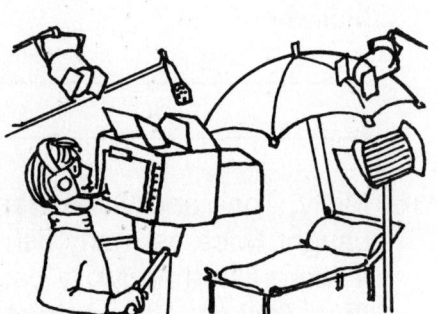

PRACTICE: Using Proportions

Write the letter of the correct proportion.

1. To serve 4 people chicken stew, Barry uses 6 tablespoons of chopped fresh basil. How many tablespoons of basil should Barry use for 12 people?

   a. $\frac{6}{12} = \frac{n}{4}$
   b. $\frac{6}{4} = \frac{n}{12}$
   c. $\frac{4}{1} = \frac{n}{12}$

2. Barry uses 1 pound of tomatoes to make chicken stew to serve to 4 people. To serve 8 people chicken stew, how many pounds of tomatoes should he use?

   a. $\frac{1}{4} = \frac{n}{8}$
   b. $\frac{1}{8} = \frac{n}{4}$
   c. $\frac{4}{8} = \frac{n}{1}$

Write a proportion when appropriate and solve.

3. Renee has a recipe for zucchini soup. She uses 3 lb of medium zucchini in the soup, which is enough for 4 people. How many pounds of zucchini would she need to make soup for 6 people?

   _____

4. Renee's zucchini soup recipe calls for $4\frac{1}{2}$ cups of milk. If this is enough for 4 people, how many cups will she need for 10 people?

   _____

5. One of Larry's dessert recipes calls for whipping cream. The cost of $\frac{1}{2}$ pint of whipping cream is $0.79. How many half-pint containers of whipping cream can he buy for $3.50?

   _____

6. Silvan can prepare a meal for 4 people in $2\frac{1}{2}$ hours. He figures that it would take him $1\frac{1}{2}$ times as long to prepare a meal for 8 people. How many hours would it take him to prepare the meal for 8 people?

   _____

7. Silvan's gourmet dinner for 4 people costs about $14.00 to prepare. About how much would a dinner for 12 people cost?

   _____

8. On Simone's TV gourmet cook show, the cameras show 3 minutes of TV cooking time for every 12 minutes of real cooking time. If a recipe takes 5 minutes of TV cooking time, how long will it take in real cooking time?

   _____

Use with pages 274–275.

PRACTICE: Scale Drawings

This is a scale drawing of one floor of the movie-studio health club.

Scale: 1 cm = 4 m

Measure the drawing and find the actual dimensions of the rooms.

1. Reception _____
2. Gymnasium B _____
3. Swimming Pool _____
4. Locker Room _____
5. Gymnasium A _____
6. Equipment Closet _____
7. Sauna _____
8. Aerobic-Dance Studio _____

A scale drawing is to be made of another floor of the health club. The scale that will be used is 2 cm = 5 m. What dimensions should the following rooms have on the new scale drawing?

9. Juice Bar (10 m × 15 m) _____
10. Lounge (15 m × 20 m) _____
11. Tennis Courts (100 m × 75 m) _____
12. Weight Room (10 m × 10 m) _____
13. Staff Room (10 m × 5 m) _____
14. Office (20 m × 5 m) _____

PRACTICE: Using a Scale Drawing

Frank Anderson designs scale models for architects. Below is the actual model of the lower level of the Hoddington Train Station.

Scale: 36 ft = 1 in.

[Scale drawing showing: Ticket office on left side; Main waiting area at top; Stairwell to main floor ($\frac{7}{8}$ in. high) on upper right; Stairwell to tracks ($\frac{3}{4}$ in. high) in middle; Magazines at bottom left; Telephones at bottom right; Waiting area on right side.]

Use the scale model to choose the correct answer. Write the letter of the correct answer.

1. What is the actual width of the station?

   a. $2\frac{1}{4}$ in.   b. 36 ft   c. 81 ft

2. What is the actual height of the stairwell to the main floor?

   a. 27 ft   b. $31\frac{1}{2}$ ft   c. 162 ft

Use the scale model to solve.

3. What is the actual length and width of the main waiting area?

   _____

4. What are the actual floor dimensions of the ticket office?

   _____

5. Which is greater, the actual height of the stairs to the main floor or the stairs leading to the railroad tracks?

   _____

6. The designer adds an information booth 9 ft long, 4.5 ft wide, and 13.5 ft high. What are the scale dimensions of the new booth?

   _____

7. How far is it from the telephones to the magazine stand in the actual station? How far is it from the telephones to the stairs leading to the main floor?

   _____

Use with pages 278–279.

113

PRACTICE — Percent

Write the percent for each ratio.

1. 6 to 100 = _____
2. 13 to 100 = _____
3. 86 to 100 = _____
4. 50 to 100 = _____
5. 1 to 100 = _____
6. 24 to 100 = _____
7. 30 to 100 = _____
8. 99 to 100 = _____
9. 39 to 100 = _____
10. 46:100 = _____
11. 33:100 = _____
12. 69:100 = _____
13. 21:100 = _____
14. 49:100 = _____
15. 17:100 = _____

Write the percent for the shaded area.

16. 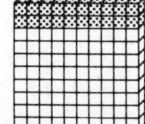 = _____

17. 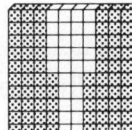 = _____

Write the percent.

18. $\frac{46}{100}$ = _____
19. $\frac{33}{100}$ = _____
20. $\frac{69}{100}$ = _____
21. $\frac{21}{100}$ = _____
22. $\frac{49}{100}$ = _____
23. $\frac{17}{100}$ = _____
24. $\frac{68}{100}$ = _____
25. $\frac{26}{100}$ = _____
26. $\frac{38}{100}$ = _____
27. $\frac{11}{100}$ = _____
28. $\frac{47}{100}$ = _____
29. $\frac{1}{100}$ = _____
30. $\frac{44}{100}$ = _____
31. $\frac{91}{100}$ = _____
32. $\frac{55}{100}$ = _____

33. 0.18 = _____
34. 0.55 = _____
35. 0.39 = _____

Solve.

36. Station WMID asked 100 people what their favorite television program was. There were 38 people who said *Valley Blues*. What percent of the people said *Valley Blues*? _____

37. In another survey of 100 people, 53 said they watch sports on television. What percent of the people do not watch sports? _____

PRACTICE: Fractions and Decimals for Percents

Write each percent as a fraction in simplest form.

1. 19% = _____
2. 76% = _____
3. 4% = _____
4. 81% = _____
5. 61% = _____
6. 97% = _____
7. 1% = _____
8. 50% = _____
9. 13% = _____
10. 80% = _____
11. 33% = _____
12. 20% = _____
13. 99% = _____
14. 5% = _____
15. 45% = _____
16. 71% = _____
17. 16% = _____
18. 29% = _____
19. 10% = _____
20. 3% = _____

Write each percent as a decimal.

21. 87% = _____
22. 41% = _____
23. 9% = _____
24. 50% = _____
25. 35% = _____
26. 98% = _____
27. 44% = _____
28. 7% = _____
29. 11% = _____
30. 99% = _____
31. 30% = _____
32. 82% = _____
33. 1% = _____
34. 73% = _____
35. 22% = _____

Solve.

36. The Record Hut ordered 100 new albums for its stock. Of these, 35% are movie sound tracks. Write a fraction in simplest form for the albums that are movie sound tracks.

37. The Record Hut estimates that 79% of their customers use charge cards. Write a decimal for the percent of customers who do not use charge cards.

Use with pages 282–283.

PRACTICE: Percents for Fractions

Write a percent for each fraction.

1. $\frac{3}{4}$ _____
2. $\frac{1}{5}$ _____
3. $\frac{5}{8}$ _____

4. $\frac{7}{20}$ _____
5. $\frac{3}{7}$ _____
6. $\frac{4}{9}$ _____

7. $\frac{9}{25}$ _____
8. $\frac{5}{6}$ _____
9. $\frac{23}{50}$ _____

10. $\frac{9}{10}$ _____
11. $\frac{8}{15}$ _____
12. $\frac{1}{3}$ _____

13. $\frac{18}{25}$ _____
14. $\frac{21}{40}$ _____
15. $\frac{17}{20}$ _____

16. $\frac{8}{30}$ _____
17. $\frac{24}{25}$ _____
18. $\frac{54}{72}$ _____

19. $\frac{11}{12}$ _____
20. $\frac{29}{100}$ _____
21. $\frac{3}{16}$ _____

22. $\frac{5}{13}$ _____
23. $\frac{8}{9}$ _____
24. $\frac{47}{50}$ _____

Write each fraction as a percent. The answers will form a magic square. Find the magic sum. _____

$\frac{3}{20}$	$\frac{3}{50}$	$\frac{3}{25}$	$\frac{1}{100}$
$\frac{1}{10}$	$\frac{3}{100}$	$\frac{13}{100}$	$\frac{2}{25}$
$\frac{1}{20}$	$\frac{4}{25}$	$\frac{1}{50}$	$\frac{11}{100}$
$\frac{1}{25}$	$\frac{9}{100}$	$\frac{7}{100}$	$\frac{7}{50}$

Use with pages 284–285.

PRACTICE: Percent of a Number

Complete.

1. 25% of 44 = _____
2. 15% of 120 = _____
3. 60% of 85 = _____
4. 16% of 36 = _____
5. 94% of 82 = _____
6. 7% of 49 = _____
7. 10% of 216 = _____
8. 11% of 78 = _____
9. 23% of 200 = _____
10. 82% of 50 = _____
11. 75% of 180 = _____
12. 55% of 125 = _____

Find the amount of each discount to the nearest cent. Then find each sale price.

13. Price: $15.00

    Discount: 20%

    Amount of Discount: _____

    Sale Price: _____

14. Price: $29.95

    Discount: 15%

    Amount of Discount: _____

    Sale Price: _____

15. Price: $250.00

    Discount: 25%

    Amount of Discount: _____

    Sale Price: _____

Copy and complete the table.
Use a sales tax rate of 6%. Round to the nearest cent.

	Item	Price	Sales tax	Total price
16.	Camera	$350.00		
17.	Photo album	$7.95		
18.	Carrying case	$39.50		
19.	Telephoto lens	$120.00		
20.	Tripod	$45.99		

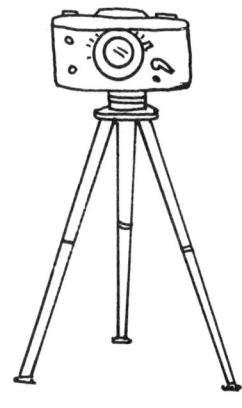

Use with pages 286–287.

PRACTICE Finding Percents

Solve.

1. What percent of 132 is 66? _____
2. What percent of 80 is 8? _____
3. What percent of 72 is 18? _____
4. What percent of 50 is 32? _____
5. What percent of 850 is 51? _____
6. What percent of 50 is 12? _____
7. 35 is what percent of 50? _____
8. 2 is what percent of 10? _____
9. 27 is what percent of 30? _____
10. 44 is what percent of 88? _____
11. 29.9 is what percent of 65? _____
12. 7.84 is what percent of 56? _____
13. 48 is what percent of 75? _____
14. 3.45 is what percent of 69? _____
15. 3.96 is what percent of 99? _____
16. 21.30 is what percent of 71? _____
17. What percent of 100 is 98? _____
18. What percent of 100 is 8? _____
19. What percent of 6 is 6? _____
20. What percent of 950 is 51? _____
21. What percent of 25 is 16? _____
22. What percent of 135 is 135? _____

Alonso's Art Supply Store is having a big sale. Find the rate of discount for each sale item.

Original price—Sale price   Discount

23. Paint set

24. Construction paper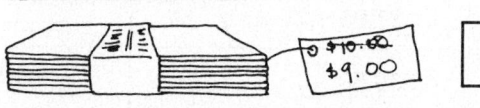

25. Scissors

26. Easel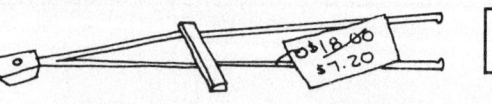

27. Modeling clay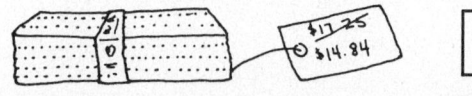

PRACTICE Finding the Total Number

Solve.

1. 50% of what number is 33? _____
2. 70% of what number is 49? _____
3. 12% of what number is 24? _____
4. 40% of what number is 10? _____
5. 20% of what number is 6? _____
6. 18% of what number is 54? _____
7. 25% of what number is 13? _____
8. 1% of what number is 5? _____
9. 45% of what number is 18? _____
10. 32% of what number is 8? _____
11. 37% of what number is 37? _____
12. 20% of what number is 8? _____
13. 9 is 2% of what number? _____
14. 21 is 28% of what number? _____
15. 16 is 20% of what number? _____
16. 45 is 30% of what number? _____
17. 140 is 35% of what number? _____
18. 6 is 8% of what number? _____

Solve.

19. There was a photography exhibit and sale at the County Center. By the time the exhibit closed, 46 black-and-white photographs and 52 color photographs had been sold. Only 30% of all the photographs in the exhibit were not sold. How many photographs were there in the exhibit?

20. Later that same month, the County Center held an exhibit of paintings and watercolors. By the end of the exhibit, 28 oil paintings and 68 watercolors were sold. In all, 80% of the paintings and watercolors were sold. How many paintings and watercolors were there in the exhibit?

Use with pages 290–291.

PRACTICE: Using a Circle Graph

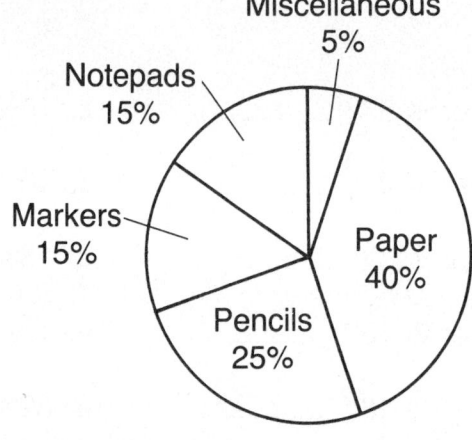

THE SCHOOL STORE INVENTORY

Use the information in the circle graph to answer each question.

1. Which item makes up most of the store's inventory? _____

2. What percentage of the inventory is made up of items other than paper? _____

3. The store's inventory is worth $230. How much are the markers worth? _____

4. What percentage of the inventory is made up of erasers? _____

5. Lined paper makes up 60% of the paper inventory. What percentage of the store's inventory is made up of lined paper? _____

6. The store stocks twice as many large notepads as small notepads. What percentage of the inventory is made up of large notepads? _____

7. Does the store stock more paper or more pencils and markers? _____

8. Do paper and pencils make up more than or less than half of the inventory? _____

9. If miscellaneous items do not include any paper products, what percentage of the inventory includes nonpaper items? _____

10. The store's inventory is worth $230. What is the value of all the nonpaper items? _____

11. The School Store makes a 30% profit on everything it sells. The entire inventory cost $345 to stock. How much profit would be made if the entire inventory was sold? _____

12. A 26% profit is made on all paper products. If the entire inventory is worth $467, how much profit would be made if all the paper products were sold? _____

PRACTICE: Using a Recipe

MARLA'S HEALTH SHAKE

- 1 cup milk
- 1 tablespoon cocoa
- $\frac{1}{8}$ teaspoon vanilla
- $\frac{1}{2}$ large banana
- 1 cup ice (ice cubes are fine)

Place milk, cocoa, vanilla, and banana into blender. Blend. Add ice and blend until smooth. Makes two 6-ounce servings. About 80 calories.

Use the information in the recipe to solve.

1. Marla decides to make four 6-ounce servings. How much vanilla should she use?

2. If Marla doubles the recipe, will the number of calories per serving double?

3. Marla figures that 8 ice cubes equal 1 cup of ice. How many ice cubes will she need if she doubles the recipe?

4. Marla has 1 pint of milk. Does she have enough milk to triple the recipe?

5. Marla estimates that it costs about $0.47 to make the recipe. About how much will it cost to triple the recipe?

6. Marla bought 5 bananas. How many batches of the recipe can Marla make with the bananas?

7. Marla wants to serve four 8-oz servings of the shake. How many teaspoons of vanilla will she need?

8. Marla wants to make $2\frac{1}{2}$ batches of the shake. Will 1 quart of milk be enough?

9. Marla throws a party for 15 people. How many bananas did she buy to make shakes for her guests?

10. How many servings can Marla make with $1\frac{1}{2}$ quarts of milk?

11. It costs $0.60 to make two servings of shake. Will $2.50 be enough to buy ingredients for $4\frac{1}{2}$ servings?

12. Marla can buy a bag of ice that holds about $4\frac{1}{2}$ cups of ice. How many bags of ice will she need for her party of 15?

Use with pages 294–295.

PRACTICE: Basic Vocabulary of Geometry

Use the figure at the right to answer Exercises 1–6.

1. Name four points on $\overleftrightarrow{JK}$. _____

2. Name two line segments with endpoint M. _____

3. Name three line segments on $\overrightarrow{JN}$. _____

4. Name an angle with $\overrightarrow{JM}$ as a side. _____

5. Name two angles with vertex N. _____

6. Name three rays with endpoint P. _____

Draw a figure for each.

7. ∠XYZ   8. $\overline{DE}$ 9. plane m

10. $\overleftrightarrow{GH}$ 11. $\overrightarrow{ST}$ 12. point Q

13. Trace the five points below. Draw as many line segments as you can joining these points. How many are there? _____

122   Use with pages 306–307.

PRACTICE — Angles

Use the figure at the right to answer Exercises 1–4.

1. Name three right angles.
   _____

2. Name three acute angles.
   _____

3. Name an obtuse angle. _____

4. Name a straight angle. _____

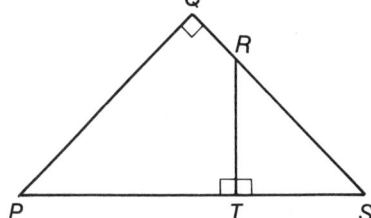

Measure the following angles with a protractor.

5.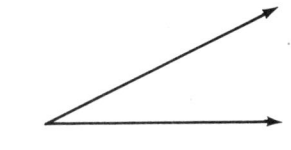

6.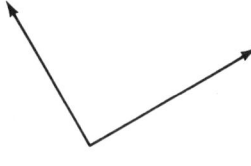

7.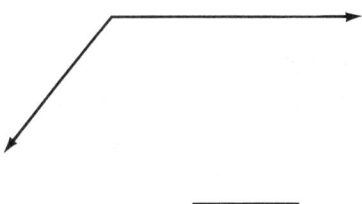

8.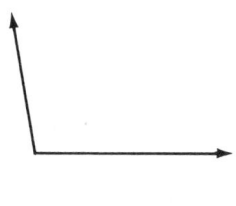

Use a protractor to draw each angle.

9. 100°   10. 50°   11. 180°

12. 75°   13. 10°   14. 125°

Use with pages 308–309.

PRACTICE: Perpendicular and Parallel Lines

Write *parallel* or *intersecting* to describe each pair of lines.

1.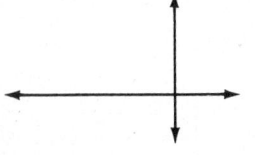
   _____

2.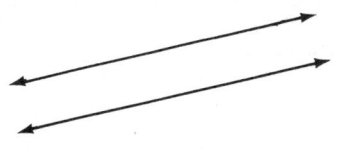
   _____

3.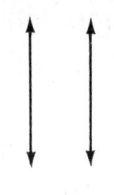
   _____

Write *perpendicular* or *not perpendicular* to describe each pair of lines.

4.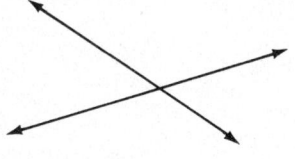
   _____

5.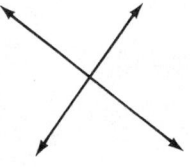
   _____

6.
   _____

Use the figure at the right to answer Exercises 7–10.

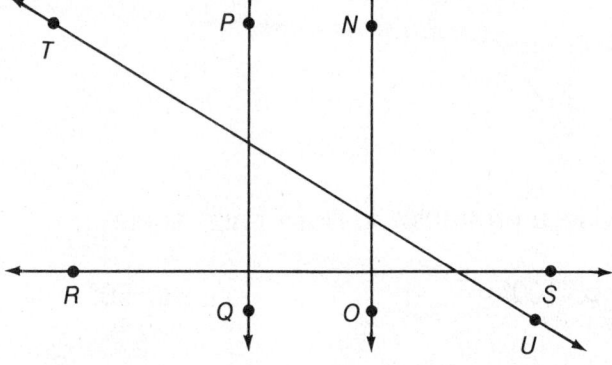

7. Name two pairs of perpendicular lines.
   _____

8. Name a pair of parallel lines. _____

9. Name a line that intersects $\overleftrightarrow{PQ}$ but is not perpendicular to $\overleftrightarrow{PQ}$. _____

10. Name a line that intersects $\overleftrightarrow{RS}$ but is not perpendicular to $\overleftrightarrow{RS}$. _____

11. Trace the line below. Construct a line perpendicular to this line.

    ←————•————→
         D

PRACTICE — Drawing a Picture

Write the letter of the drawing you would use to solve each problem.

1. Karen has arrived at camp and is following written directions to her cabin. She walks through the entrance to the camp grounds and walks 25 paces straight ahead. Then, she turns right and walks 15 paces. Then, she turns left and walks 20 paces. Finally, she turns right and walks 30 paces to her cabin. If each of her paces equals one yard, how far is Karen's cabin from the campground entrance?

2. Mr. Dawson put up a fence on the border of his rectangular property. He installed 50 feet of fencing on the north side, 75 feet on the east side, 50 feet on the south side, and 75 feet on the west side. How much fencing did Mr. Dawson use?

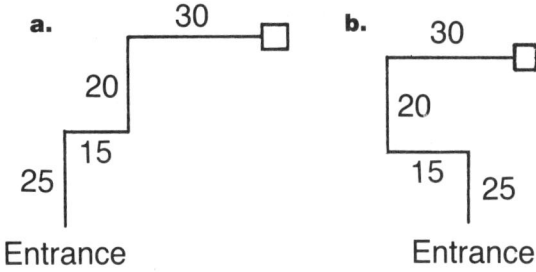

Solve. Draw a picture if you need to.

3. There are 5 houses on Queen Street. The distance between each house is 30 feet, and the end houses are 20 feet from the end of the street. The length in feet of the front of each house is 32, 35, 31, 36, and 38. How long is the street?

4. Mr. Dawson's mailbox is 20 feet from his house. In front of the mailbox is a sidewalk 3 feet wide. The sidewalk is 6 feet from the road. How far is Mr. Dawson's house from the road?

5. Mrs. Young walks out the front door of her house to look for her kitten. Her house is 42 ft long and 28 ft wide, and the front door is located in the middle of the wall that faces the street. She walks around the house to the back door, where she finds her kitten. How far did Mrs. Young walk?

6. Johnny is walking to Wally's house. Johnny passes Cindy's house, which is 8 houses from his house. Ray's house is 3 houses before Cindy's, and Wally's house is 10 houses past Ray's. How many houses is it from Johnny's house to Wally's house?

Use with pages 312–313.

PRACTICE   Triangles

Name each triangle according to the lengths of its sides.

1.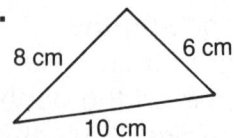
_____

2.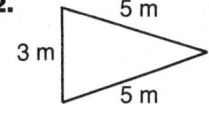
_____

3.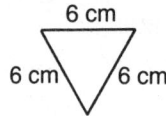
_____

4.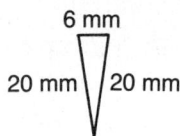
_____

Name each triangle according to the measure of its angles.

5.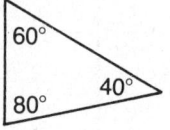
_____

6.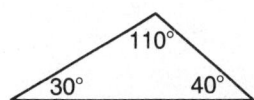
_____

7.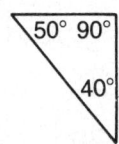
_____

8.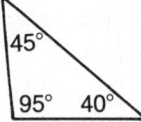
_____

Find the measure of the missing angle.

9.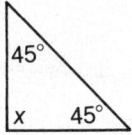
_____

10.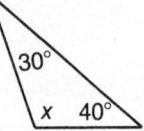
_____

11.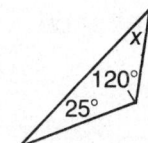
_____

12.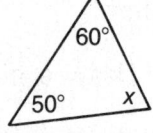
_____

Look at the figure at the right.

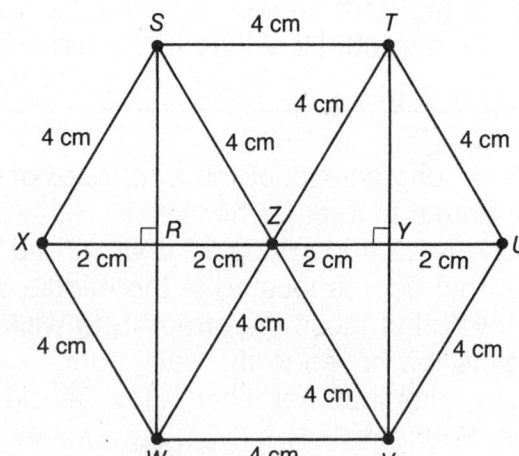

13. Count and name all the equilateral triangles.
_____

14. Count and name all the right triangles.
_____
_____

15. Count and name all the isosceles triangles.
_____
_____

126

Use with pages 314–315.

PRACTICE Problem-Solving Practice

The members of the King School Camping Club are planning their program for the school year. They have $200 to spend. The circle graph shows how they plan to spend their money.

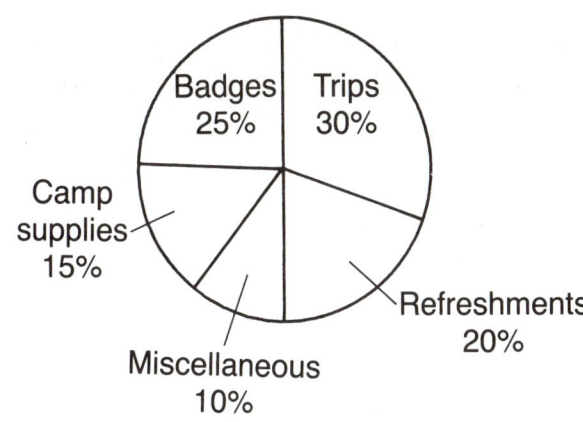

Write the letter of the operation you would use to solve each problem.

1. Will the club spend twice as much on trips as on supplies?
   a. add
   b. subtract
   c. multiply
   d. divide

2. How much more will the club spend on badges than on refreshments?
   a. add
   b. subtract
   c. multiply
   d. divide

Make a plan to solve each problem.

3. Each badge costs $1.25. How many badges can the club buy?

   Step 1: _____
   Step 2: _____
   Step 3: _____

4. If the club spent 25% of their trip funds, how much money would be left for trips?

   Step 1: _____
   Step 2: _____
   Step 3: _____

Decide whether you would estimate or find the exact answer to solve. Write *estimate* or *exact*.

5. The club spent $12 on refreshments. How much money is left for refreshments? _____

6. About what fraction of the club's money is spent on trips and camp supplies? _____

Use with pages 316–317.

PRACTICE Problem-Solving Practice

Write the letter of the best method of solving each problem.

7. There are 45 students going on a field trip. Each bus holds 18 students. How many buses will be needed?

   a. round quotient up
   b. round quotient down
   c. use remainder only

8. The club plans a canoe trip for 27 students. Each canoe holds 4 people. How many canoes will be full?

   a. round quotient up
   b. round quotient down
   c. use remainder only

Solve.

9. The club plans to buy 5 cases of juice for the school year. Each case costs $4.00. How much will be left in the club's budget of $200 to spend on other kinds of refreshments? _____

10. The club decides to take $10 from the miscellaneous fund of the budget for refreshments. What percent of the total budget will now be spent on refreshments? _____

11. If $11 of the fund for supplies has already been spent, is there enough money left in that fund to buy 3 flashlights that sell for $5 each?
_____

12. If the club raises $50 and adds it to the trip fund, what percent of the total funds can the club spend on trips?
_____

13. Club members are planning a 17.1-mile hike. If they hike 3.8 miles per hour, how many full hours will the hike take?
_____

14. If the club spends 75% of the funds for trips and badges and 80% of the funds for all other categories, how much money is left?
_____

15. If each of 34 students received 3 badges, how many badges would be left in the club's supply, which originally totaled 212 badges? _____

16. Club members are planting trees in a local park. Each sapling needs 5 square feet of ground. How many saplings can members plant in 244 square feet? _____

17. If the club's funds amounted to $225, how much would members be able to spend on refreshments? _____

18. The club receives a gift of $120. The members decide to use the gift in the same manner in which they budgeted their original $200. If they follow their plan, how much money will they budget for trips? _____

PRACTICE Quadrilaterals

Write the name that best describes each figure.

1.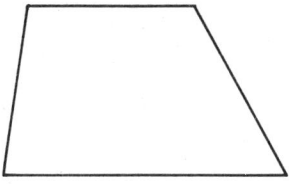

2.

3.

4.

5.

6.

7.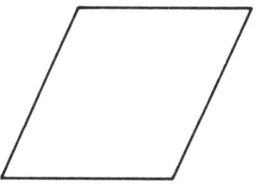

8.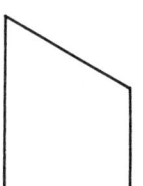

9.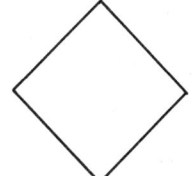

Use the figures to answer Exercises 10–12.

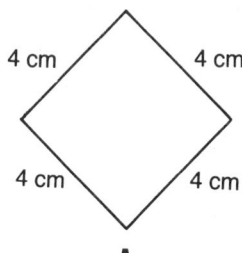
A

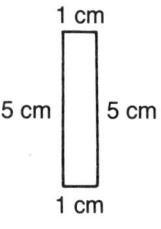
B

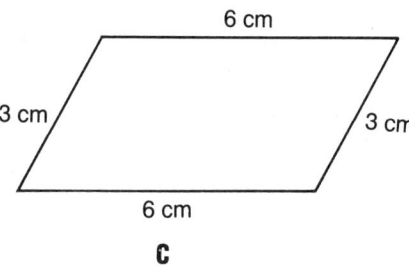
C

10. Which of the figures are parallelograms? _____

11. Which of the figures are rectangles? _____

12. Which of the figures are rhombuses? _____

Use with pages 318–319.

PRACTICE Other Polygons

Write the name of each polygon.

1.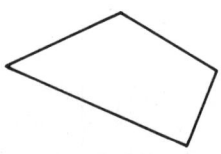
_____

2.
_____

3.
_____

4.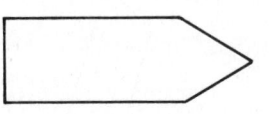
_____

5.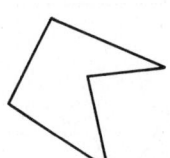
_____

6.
_____

Find the sum of the angles of each figure.

7. a quadrilateral
_____

8. a hexagon
_____

Trace the figure at the right to answer Exercises 9–11.

9. Draw all the diagonals you can from vertex A.

10. How many triangles are formed by the triangles you drew? _____

11. Find the sum of the angles of the polygon. _____

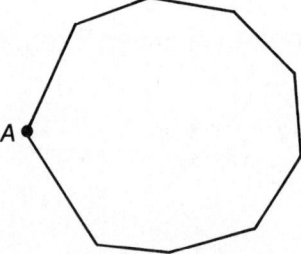

PRACTICE: Congruence and Symmetry

Write the letter of each polygon that is congruent to the given polygon.

1.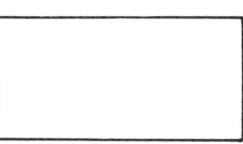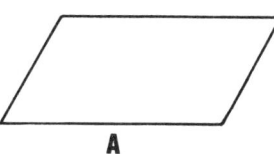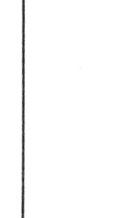
   A   D

2.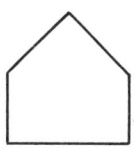
   B

3.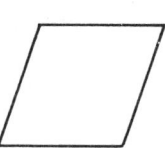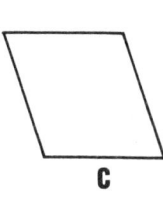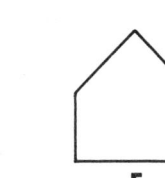
   C   E   F

Triangle PQR ≅ triangle XYZ. Copy and complete the chart.

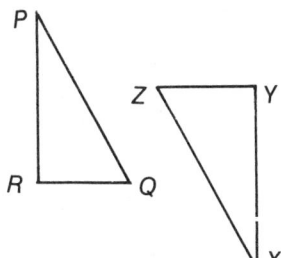

4.	$\overline{PQ} \cong$ _____	$\angle P \cong$ _____
5.	$\overline{PR} \cong$ _____	$\angle Q \cong$ _____
6.	$\overline{RQ} \cong$ _____	$\angle R \cong$ _____

Trace each figure, then draw all the possible lines of symmetry.

7.    8. 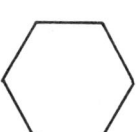   9.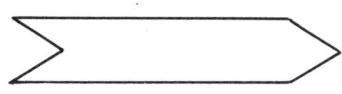

A figure has exactly three lines of symmetry. The lines of symmetry and part of the figure are shown. Copy and complete the figure.

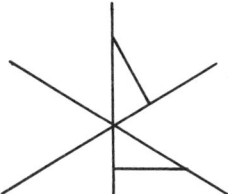

Use with pages 322–323.

131

PRACTICE   Similar Figures

Write the letter of the figure that is similar to the given figure.

1.  _____

2. 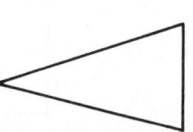 _____

3. 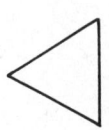 _____

4. 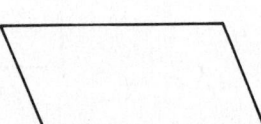 _____

a.

b.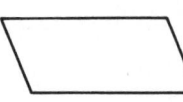

c.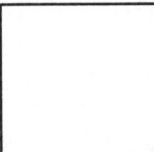

d.

Figure *JKLM* ~ figure *NOPQ*. Copy the chart below. Name the corresponding parts.

5.	$\overline{JK}$ ~ _____	∠J ≈ _____
6.	$\overline{KL}$ ~ _____	∠K ≈ _____
7.	$\overline{LM}$ ~ _____	∠L ≈ _____
8.	$\overline{MJ}$ ~ _____	∠M ≈ _____

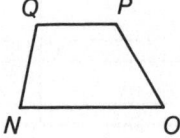

Figure *ABCDE* ~ figure *STUVW*. Find the measure of the angles and lengths of the sides.

9. measure of ∠B = _____

10. measure of ∠U = _____

11. measure of ∠S = _____

12. length of $\overline{ST}$ = _____

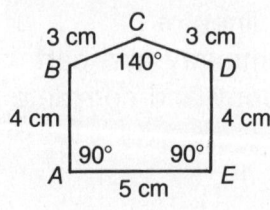
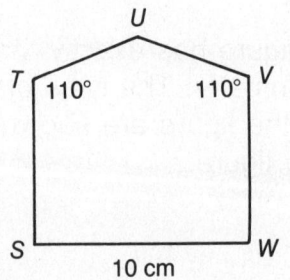

Use with pages 324–325.

PRACTICE Circle

Write *chord*, *diameter*, or *radius* for each line segment.

1.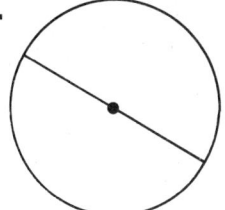

2.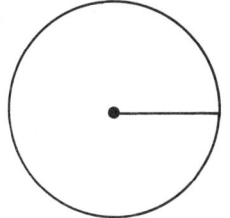

3.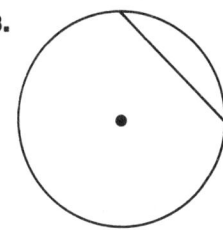

4.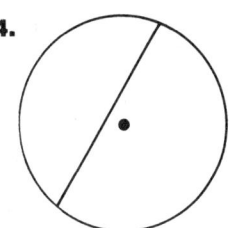

_____   _____   _____   _____

Use the circle to answer Exercises 5–9.

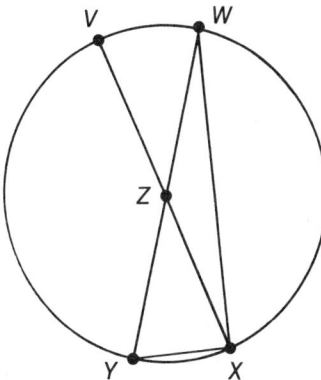

5. Name the circle. _____

6. Name two diameters. _____

7. Name two chords. _____

8. Name an acute central angle. _____

9. Name four radii. _____

Solve.

10. Circle P is congruent to circle Q. Circle P has a radius of 11 cm. What is the diameter of circle Q? _____

11. Trace $\overline{AB}$. Construct a circle with center A that has $\overline{AB}$ as a radius.

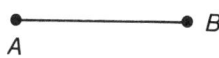

Use with pages 326–327.

PRACTICE: Translations, Rotations, Reflections

Describe the relationship of each pair of polygons. Write *translation, rotation,* or *reflection.*

1.

2.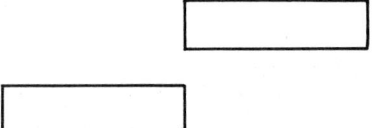

3.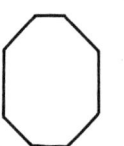

4.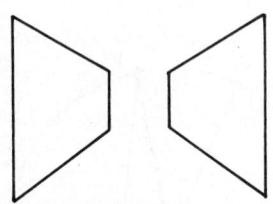

5.

6.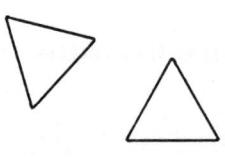

Trace each figure. Draw each figure's reflection on the line of symmetry to make a symmetrical figure.

7.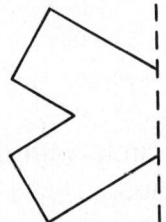

8.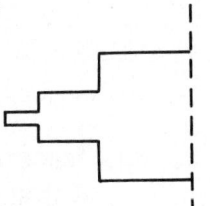

9.

Trace each figure and point *P*. Draw the image of each figure after it has been rotated a half turn around point *P*.

10.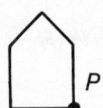

11.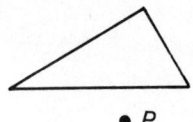

PRACTICE: Looking for a Pattern

Solve by finding the pattern.

1. Marcie uses a secret code to write in her diary. She has written one key word to a six-word sentence on each of the following pages: 1, 3, 6, 10, and 15. On which page is the last word of the sentence found? _____

2. The Metro bus stops at the shopping mall at the following times: 9:00 A.M., 10:15 A.M., 11:30 A.M., and 12:45 P.M. At what time will the sixth stop be made at the shopping mall? _____

3. Joe is measuring the distance between rows in his father's garden. The second row is 2 inches from the first row; the third row is 6 inches from the second; the fourth row is 12 inches from the third; and the fifth row is 20 inches from the fourth. What pattern did Joe's father follow in planting the rows in his garden? _____

4. Amy is putting pennies in piles. The piles of pennies are to be used for prizes to be awarded at the school carnival. After putting 1 penny in the first pile, 4 pennies in the second pile, 9 pennies in the third pile, and 16 pennies in the fourth pile, she is following the pattern as she completes the job. How many pennies did she put in the fifth pile? the eighth pile? _____

5. While Julia was looking in a math book in the library, she saw a strange decimal: 2.141141114. . . . She asked her teacher what the three dots at the end of the number meant. Her teacher said the three dots meant to continue the same pattern. Can you see the pattern in the digits used? Write the next decimal in the pattern. _____

6. Manny writes a code on his coin holders to remind him how much he paid for each coin in his collection. The letter A means he paid 15¢, the letter B means he paid 28¢, the letter C means he paid 41¢, and the letter D means he paid 54¢. How much did Manny pay for a coin on whose holder he had written the letter G? _____

7. Carla counted the number of students in her homeroom. There are six rows of desks in the classroom and five desks in a row. Sammy sits in the fifth desk of the first row. Carla started with the first desk. What number did Carla use to count Sammy? _____

8. How many rectangles are there in this arrangement? What is the pattern?

a. ____ b. ____ c. ____ d. ____

_____

Use with pages 330–331.

PRACTICE: Looking for a Pattern

Solve by finding the pattern.

9. Mary Ann is working on some puzzles she found in a school newspaper. She is trying to find the answer to 1111 × 1111 without actually multiplying the two numbers. Mary Ann knows that 1 × 1 = 1 and 11 × 11 = 121. She tries 111 × 111. The answer is 12,321. Now she knows the answer to 1111 × 1111 without multiplying. What is the answer?

10. Mrs. Rae will hold a luncheon for her friends. She plans to serve tea. She noticed that 1 tea bag can make 2 cups of tea, 2 tea bags can make 5 cups of tea, and 3 tea bags can make 9 cups of tea. At this rate, how many cups of tea can Mrs. Rae make with 6 tea bags?

11. Thomas learned how to do push-ups in gym class. They were difficult to do at first, but after a few days, Thomas was able to do several without stopping. On the first day, he was able to do only 1 push-up. On the second day, he did 5 push-ups without stopping. On the third day, he did 14 without stopping. At this rate, how many push-ups did Thomas do on the fifth day?

12. Carlo and Andrew painted numbers on school lockers. Carlo painted odd numbers on the lockers located on the right side of the hallway. He painted a 1 on the first locker. On the second locker, he painted a 3. On the third locker, he painted a 5. What number did he paint on the twenty-fifth locker?

13. Dr. Jackson is testing different processor chips for the Science Club. A 1K chip has a processing speed of 32 milliseconds. A 3K chip has a processing speed of 16 milliseconds. What is the processing speed of a 5K chip?

PRACTICE: Perimeter of Polygons

Write the perimeter of each polygon.

1. 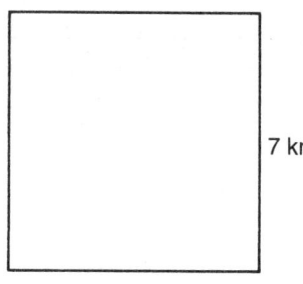 7 km

2. 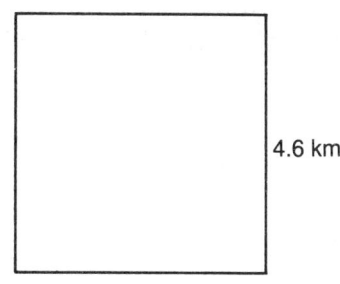 4.6 km

3. 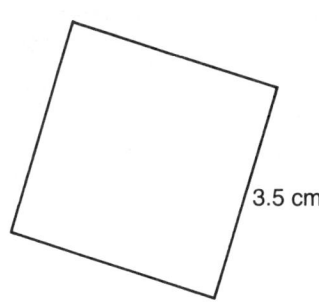 3.5 cm

4. 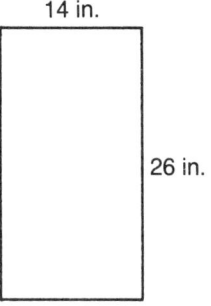 14 in., 26 in.

5. 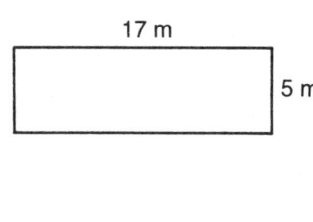 17 m, 5 m

6. 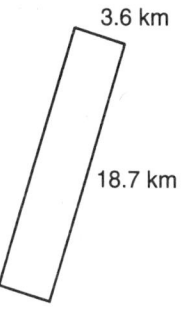 3.6 km, 18.7 km

7. 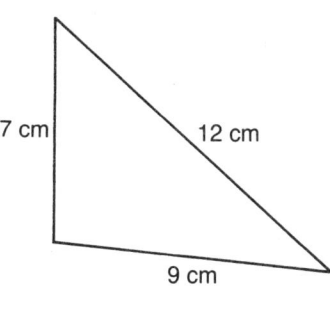 7 cm, 12 cm, 9 cm

8. 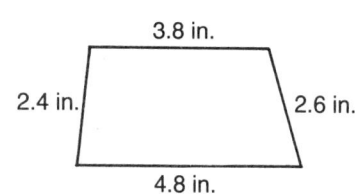 3.8 in., 2.4 in., 2.6 in., 4.8 in.

9. 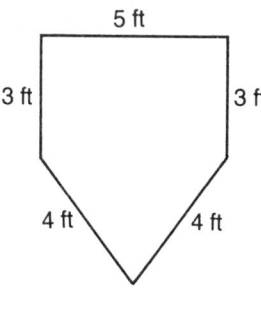 5 ft, 3 ft, 3 ft, 4 ft, 4 ft

10. 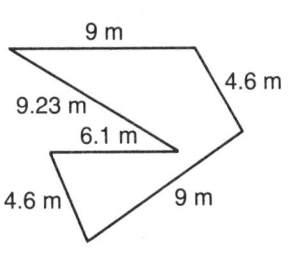 9 m, 4.6 m, 9.23 m, 6.1 m, 4.6 m, 9 m

11. 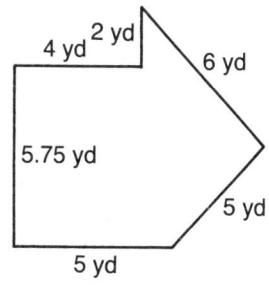 4 yd, 2 yd, 6 yd, 5.75 yd, 5 yd, 5 yd

12. 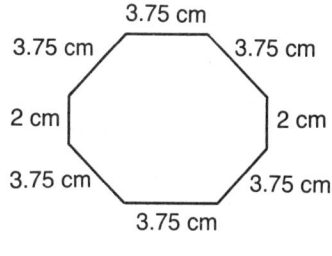 3.75 cm, 3.75 cm, 3.75 cm, 2 cm, 2 cm, 3.75 cm, 3.75 cm, 3.75 cm

Use with pages 342–343.

PRACTICE: Using Line and Bar Graphs

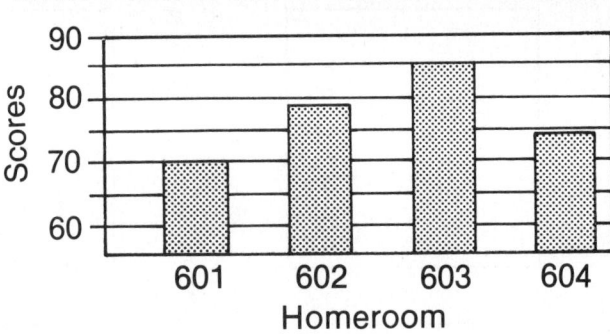

Use the line graph and the bar graph to answer each question.

1. How many quizzes has Mark taken? _____

2. What was Mark's lowest score? _____

3. What was Mark's highest score? _____

4. Was Mark's third quiz score higher or lower than his first quiz score? _____

5. After Mark received his lowest score, did his scores improve? _____

6. How many points higher is Mark's highest score than his lowest score? _____

7. What is the average of Mark's quiz scores? _____

8. Mark is in one of Mrs. Port's math classes. Which homeroom do you think Mark is in? _____

9. What would Mark's average score be if his first quiz score had been 10 points higher? _____

10. Which of the sixth-grade homerooms had the highest-average quiz score? _____

11. What is the difference between the highest-average quiz score and the lowest-average quiz score? _____

12. If your average quiz score is 72, which homeroom would you probably be in? _____

13. How many sixth-grade math classes does Mrs. Port teach? _____

14. How many students in room 603 have Mrs. Port for math? _____

Use with pages 344–345.

PRACTICE: Circumference

Write the circumference. Use 3.14 for π.

1.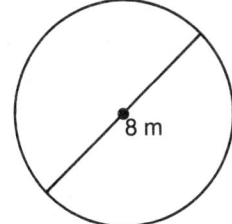
2.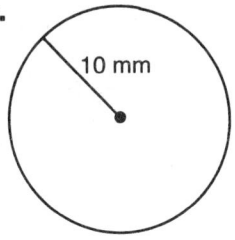
3.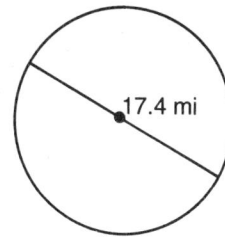
4.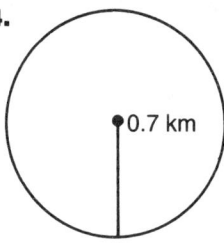

_____   _____   _____   _____

5. $r = 0.5$ m

6. _____ $r = 7.03$ ft   _____

7. $r = 0.75$ m   _____   8. $d = 0.009$ cm   _____

9. $d = 40$ yd   _____   10. $d = 26.4$ ft   _____

11. $r = 0.1$ m   _____   12. $d = 0.52$ yd   _____

Write the circumference. Use $\frac{22}{7}$ for π.

13.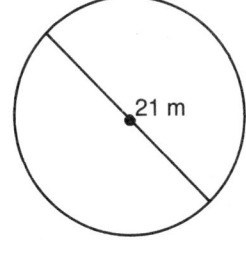
14.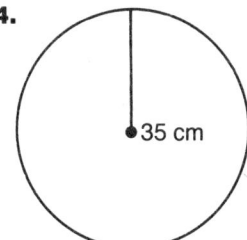
15.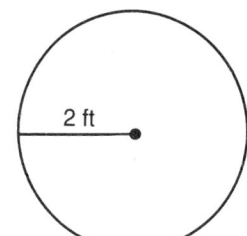
16.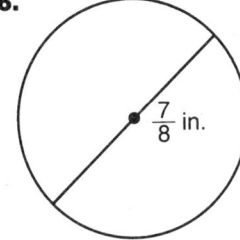

_____   _____   _____   _____

Use $\frac{22}{7}$ for π. Copy and complete the table.

17.

Diameter	7 cm	14 cm	28 cm	56 cm	84 cm
Circumference					

Use with pages 346–347.

PRACTICE: Area of Rectangles and Squares

Find the area of each square.

1. s = 5 cm _____
2. s = 10.1 in. _____
3. s = 8.04 m _____
4. s = 13.2 ft _____
5. s = 3.301 in. _____
6. s = 0.31 yd _____
7. s = 90 cm _____
8. s = 22 ft _____

Find the area of each rectangle.

9. l = 9 m
w = 7 m
A = _____

10. l = 3.33 yd
w = 3 yd
A = _____

11. l = 41 in.
w = 7 in.
A = _____

12. l = 41 ft
w = 8.4 ft
A = _____

13. l = 84 in.
w = 12 in.
A = _____

14. l = 55 cm
w = 8.2 cm
A = _____

15. l = 1 m
w = 82 m
A = _____

16. l = 1.2 ft
w = 8 ft
A = _____

17. l = 92 yd
w = 43 yd
A = _____

Find the area.

18.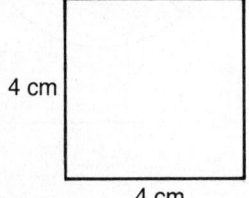
4 cm by 4 cm
_____

19.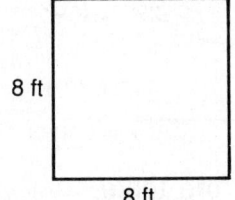
8 ft by 8 ft
_____

20.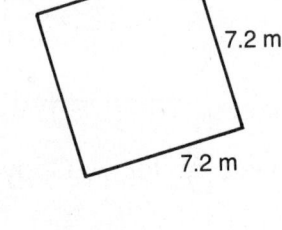
7.2 m by 7.2 m
_____

21.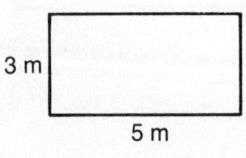
3 m by 5 m
_____

22.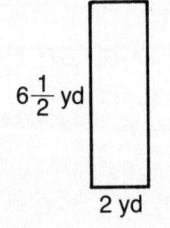
$6\frac{1}{2}$ yd by 2 yd
_____

23.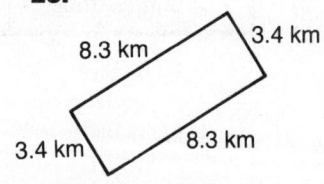
8.3 km by 3.4 km
_____

PRACTICE | Choosing and Writing a Sensible Question

Some students from Kennedy School are interested in going on a camping trip. They would like to spend a weekend in one of two large state parks, Big Run Park or Red Wood Park.

Read each statement. Write questions that should be answered before making a decision.

1. The students have to decide in which state park to spend the weekend.

   _____
   _____
   _____
   _____

2. The students have decided to go to Big Run Park. They have to decide what equipment to take.

   _____
   _____
   _____
   _____

3. The students called the state ranger to ask if they would be allowed to build a fire in the park. Her answer will help them decide which food to bring for the weekend.

   _____
   _____
   _____
   _____

4. The ranger sent the students a scale drawing of the park. The students located five possible campsites. They have to decide which one would be the best for the campsite.

   _____
   _____
   _____
   _____

Use with pages 350–351.

PRACTICE: Area of Parallelograms and Triangles

Write the area of each parallelogram.

1. b = 5 cm
   h = 12 cm
   A = _____

2. b = 14.2 cm
   h = 3 cm
   A = _____

3. b = 3 ft
   h = 14 ft
   A = _____

4. b = 100 m
   h = 0.001 m
   A = _____

5. b = 2.33 cm
   h = 200 cm
   A = _____

6. b = 12 in.
   h = 36 in.
   A = _____

7. b = 4.5 m
   h = 0.5 m
   A = _____

8. b = 3.1 cm
   h = 10.99 cm
   A = _____

9. b = 19 ft
   h = 93.79 ft
   A = _____

10.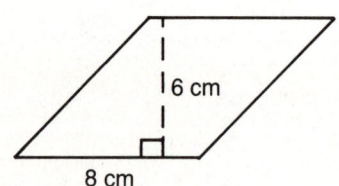
    _____

11.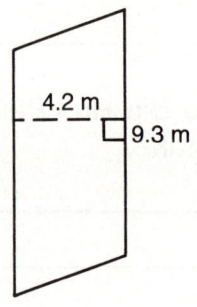
    _____

12.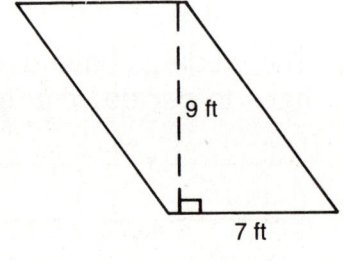
    _____

Write the area of each triangle.

13.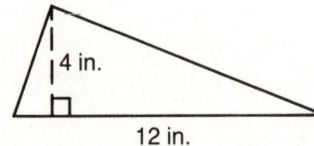
    _____

14.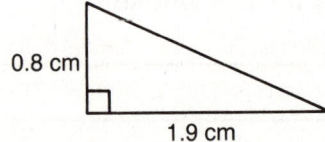
    _____

15.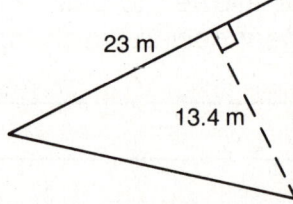
    _____

16. b = 4 cm
    h = 0.5 cm
    A = _____

17. b = 500 yd
    h = 30 yd
    A = _____

18. b = 20 ft
    h = 18 ft
    A = _____

19. b = 46.3 m
    h = 1.9 m
    A = _____

20. b = 2 in.
    h = 1.001 in.
    A = _____

21. b = 43 yd
    h = 58 yd
    A = _____

Use with pages 352–353.

PRACTICE Area of a Circle

Find the area. Use 3.14 for π.

1.
   4 cm
   _____

2.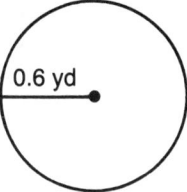
   0.6 yd
   _____

3.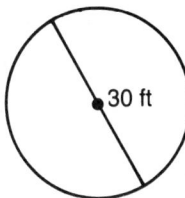
   30 ft
   _____

4.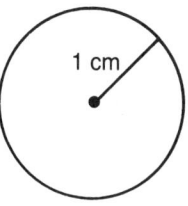
   1 cm
   _____

5.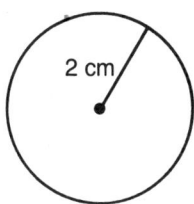
   2 cm
   _____

6.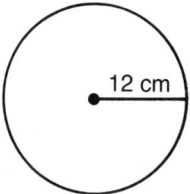
   12 cm
   _____

Find the area. Use $\frac{22}{7}$ for π.

7.
   21 km
   _____

8.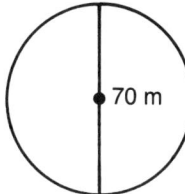
   70 m
   _____

9.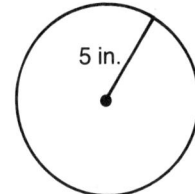
   5 in.
   _____

Find the area. Use 3.14 for π.

10. $r = 1,000$ km _____

11. $r = 1.331$ m _____

12. $r = 3$ ft _____

13. $r = 18$ in. _____

14. $r = 0.003$ cm _____

15. $r = 8$ ft _____

16. $r = 30$ km _____

17. $r = 0.5$ yd _____

18. $r = 3$ cm _____

19. $r = 6$ km _____

20. $r = 0.88$ m _____

21. $r = 46$ ft _____

Use with pages 354–355.

PRACTICE: Choosing a Formula

Write the letter of the best formula for solving the problem.

1. The art class made a banner for the school play. It was 30 ft long and 4 ft wide. How much of the wall did the banner cover?

   a. $p = 2l + 2w$
   b. $A = lw$
   c. $A = \pi r^2$

2. Corking was used to make circles on the banner. Each circle has a radius of $1\frac{1}{2}$ ft. What is the circumference of each circle?

   a. $C = 2l + 2w$
   b. $A = \pi r^2$
   c. $C = 2\pi r$

Use a formula to help you solve each problem. Solve each problem.

3. The kickball field is located on the grounds of the school. Each side measures 85 ft. What is the perimeter of the kickball field? _____

4. A large truck tire is used as a swing in the jungle gym area of the playground. The tire has a diameter of 1.8 m. What is its circumference? _____

5. A large circle has been painted on the outside basketball court. It is 4.8 m in diameter. What is the area of the circular region on the basketball court? _____

6. The school playground is rectangular in shape. The area of the playground is 2,475 m², and the side of the playground next to the street is 45 m long. How wide is the playground? _____

7. The high school has a rectangular swimming pool that is 25 m wide and 55 m long. It is filled with water. What is the surface area of the water? _____

8. One of the school buildings has circular floors. The circumference of each floor is 56.52 m. What is the area of each floor? _____

9. The floor of the handball court is 20 ft wide by 40 ft long. Three of the walls are 20 ft tall. The fourth wall is 12 ft tall and 20 ft wide. There is no ceiling. What is the surface area of the handball court? _____

10. Some students are making pennants for a forthcoming track meet. Each pennant is a triangle 6 inches wide and $1\frac{1}{2}$ ft long. How many square feet of cloth do they need to make 25 pennants? _____

Use with pages 356–357.

PRACTICE: Solid Figures

Write the name of each shape.

1.

   _____

2.

   _____

3.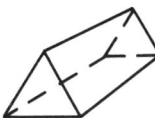

   _____

4.

   _____

5.

   _____

6.

   _____

7.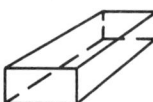

   _____

8.

   _____

9.

   _____

Copy and complete the chart.

Solid figure	Number of faces	Number of edges	Number of vertices
cube			
cylinder			
pyramid			
triangular prism			
cone			

Use with pages 358–359.

145

PRACTICE Surface Area

Write the surface area. Use 3.14 for π.

1.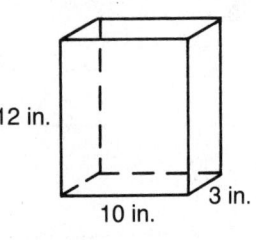

_____

2.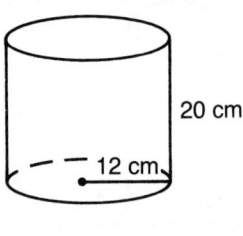

_____

3.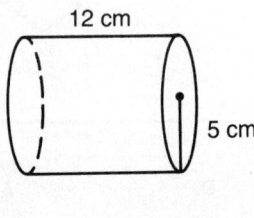

_____

4.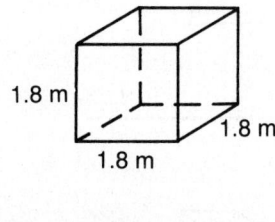

_____

5.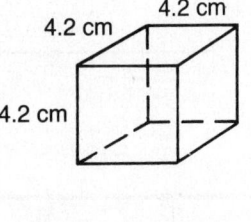

_____

6.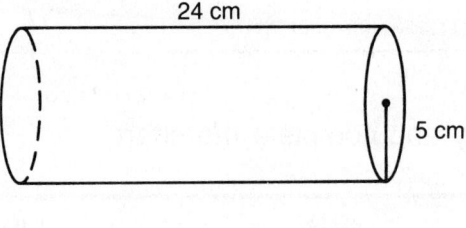

_____

7.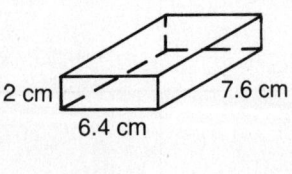

_____

8.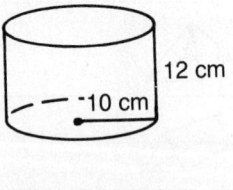

_____

PRACTICE: Volume

Write the volume.

1.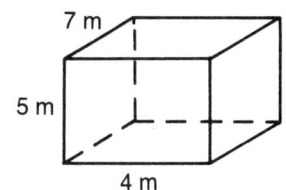

_____

2.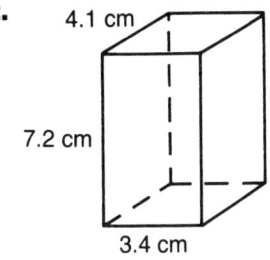

_____

3.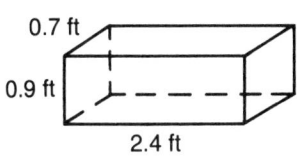

_____

4.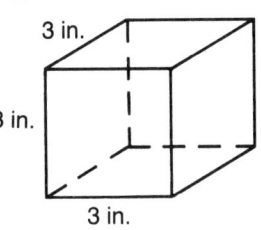

_____

5.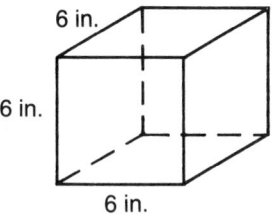

_____

6.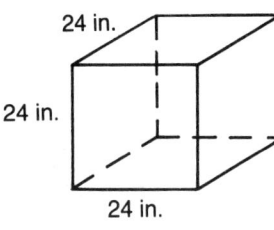

_____

7.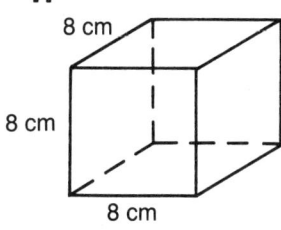

_____

8.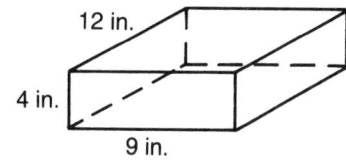

_____

9.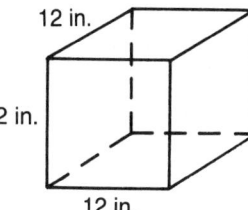

_____

10. $l = 4.3$ m
    $w = 2.2$ m
    $h = 10$ m
    $V =$ _____

11. $l = 12$ in.
    $w = 36$ in.
    $h = 17$ in.
    $V =$ _____

12. $l = 1$ yd
    $w = 2.02$ yd
    $h = 8.7$ yd
    $V =$ _____

13. $l = 6.7$ cm
    $w = 2.5$ cm
    $h = 11$ cm
    $V =$ _____

14. $l = 137$ mm
    $w = 64$ mm
    $h = 259$ mm
    $V =$ _____

15. $l = 0.5$ ft
    $w = 0.5$ ft
    $h = 0.5$ ft
    $V =$ _____

Use with pages 362–363.

PRACTICE — Drawing a Diagram

Solve. Draw a diagram if needed.

1. Students working on a class project are designing buildings for a "city of tomorrow." Celine designs an arts pavilion. The base of the building is 220 ft long, 175 ft wide, and 35 ft high. At the top is a great square pyramid. Each face of the pyramid rises 88 ft from base to vertex. Its base is 175 ft long. Celine notes on her design that the entire surface of the building is to be painted blue to match the sky. If Celine's design was used to build a pavilion, how many square feet would be painted blue? _____

2. Dejan designs a large government building. It is 166 ft high, 75 ft wide, and 60 ft deep. In front of the building is a reflecting pool. It is 75 ft wide, 166 ft long, and 5 ft deep. What is the volume of the reflecting pool? _____

3. Li designs an industrial complex. The main building is a cube measuring 1,600 yd on each side. On the roof are 4 giant cylindrical smokestacks. Each stack is 72 yd high and 12 yd in diameter. Li specifies that insulation be wrapped around the bottom half of each stack. What is the total surface area to be covered by the insulating wrap for each cylinder and for all 4 cylinders?

_____

PRACTICE Drawing a Diagram

Solve. Draw a diagram if needed.

4. Lena designs an executive-office complex. The complex is shaped like a rectangular prism. On the roof are two helicopter landing pads. The roof measures 155 ft long and is 44 ft wide. The first landing pad is 30 ft in diameter. The second pad is 27 ft in diameter. The landing pad circles are painted white. How much of the roof's surface area is not painted white?

_____

5. Velma has designed a futuristic hotel. The hotel is a 17-yd-long cube. One face rests against a rectangular tower. The tower rises to a height of 145 yd. It is 45 yd wide and 60 yd deep. Two identical wings flank the tower. The entire hotel complex is covered with unbreakable glass windows. How many square yards of glass did Velma specify to cover all the sides of the hotel?

_____

6. Robin's addition to the city is a sports arena. The arena is a hexagonal pyramid supported by 5 massive columns. Each triangle of the pyramid has a 70 ft base and rises to a height of 84.5 ft. Robin specifies that the building be covered with copper foil so that it will glisten in the sun. How many square feet of foil would be needed to cover the pyramid? _____

7. Each project has been drawn on paper 1.5 ft long and 0.75 ft wide. All the projects will be displayed on a board outside the classroom. The board is 9 ft long and 5 ft high. How many projects will fit onto the display board? _____

Use with pages 364–365.

PRACTICE: Recording Data on a Table

Most of the sixth-grade class live on Bank Street, Mill Street, or Front Street. The students have different ways to get to school.

Jack—Bank St., bike
Marcia—Mill St., walk
George—Front St., bike
Fred—Bank St., car
Susan—Mill St., walk
Jane—Front St., bus

Jim—Mill St., bike
Kelly—Front St., car
Joe—Mill St., bike
Elaine—Bank St., car
Tara—Front St., walk
Samuel—Mill St., bus

Michael—Mill St., bus
Danny—Bank St., bike
Julio—Front St., car
David—Mill St., bus
Agnes—Bank St., bike
Chris—Front St., car

Copy each table. Use the information above to complete.

1.

Transportation to school	Tally	Number of students
bus		
bike		
car		
walk		

2.

Location of home	Tally	Number of students
Bank St.		
Mill St.		
Front St.		

There are 14 students in each class. Below is a record of how many sit-ups each student was able to do. Copy the table. Use the information to complete it.

Fifth Graders: 7, 3, 12, 35, 11, 10, 28, 8, 2, 12, 19, 25, 33, 13
Sixth Graders: 5, 22, 17, 9, 20, 41, 18, 14, 22, 37, 7, 15, 10, 24
Seventh Graders: 29, 35, 12, 19, 28, 18, 6, 18, 20, 46, 14, 8, 31, 29

3.

Grade	Number of students doing:		
	0–10 sit-ups	11–20 sit-ups	more than 20 sit-ups
Fifth			
Sixth			
Seventh			
Total			

Use with pages 376–377.

PRACTICE: Interpreting Information

Copy and complete this table.

	Set	Mean	Median	Mode	Range
1.	3, 6, 6, 10, 7, 9, 15				
2.	4, 9, 13, 13, 16, 11				
3.	18, 26, 14, 27, 26				
4.	80, 77, 64, 52, 73, 89, 77, 58				
5.	7.4, 5.4, 8, 5.2, 3.1, 9.8, 3.1				
6.	53, 59, 66, 66, 70, 76, 76, 76, 79				
7.	0.3, 1.2, 8.2, 0.2, 0.9, 0.3, 0, 0.1				

The members of the sixth-grade Music Club had a sandwich sale to earn money for uniforms.

NUMBER OF SANDWICHES SOLD BY MUSIC CLUB

Club Member	Sue	Tim	Lee	Roy	Bud
Sandwiches sold	11	15	13	8	40

8. Find the mean. _____

9. Find the median. _____

10. Does the mean or the median give a more accurate idea of the number of sandwiches each member sold? Explain your answer.

_____

_____

_____

Use with pages 378–379.

PRACTICE: Making a Pictograph

Use the data in each table to make a pictograph.

1. **NUMBER OF ROBOTS IN USE (YEAR 2612)**

Planet	Number of robots
Corinna	350
Belaria	200
Nestor	650

2. **DISTANCES TRAVELED BY INTERSTELLAR FLEET**

Spaceship	Distance traveled (in parsecs)
Galaxy Queen	30
Farflight	50
Solaria	25
Corona	75
Titan	30

3. **SPACESHIP CREWS**

Spaceship	Number of people in the crew
Galaxy Queen	325
Farflight	275
Solaria	100
Corona	350
Titan	75

PRACTICE: Making a Table to Find a Pattern

Copy and complete each table to help you solve each problem.

1. The manager buys pencils for the school store. The table below shows the number of pencils bought and the price charged for them.

Number of pencils	10	50	100	500	1,000	5,000	10,000
Price	$5.00	$22.50	$40.00				

What is the price of 5,000 pencils? _____

What is the price of 10,000 pencils? _____

The largest package of pencils contains 10,000 pencils. What price would be charged for an order of 25,000 pencils? _____

2. As she trains herself for long-distance running, Sara gradually increases the number of miles she runs each day. She is keeping track of her total mileage on a chart.

Day	1	2	3	4	5	6	7
Total mileage	0.5	1.5	3.0				

How many miles had she run by the end of day 5? _____

How many miles had she run by the end of day 7? _____

The longest distance she wants to run in a day is 8 miles. Following her present schedule, on which day will she actually run 8 miles? _____

Make a table and solve.

3. Svea starts to play basketball at 10:45 A.M. and shoots about 8 baskets per minute. What time will it be by the time she shoots 48 baskets? _____

PRACTICE: Bar Graphs

Use the data in each table to make your own bar graph.

1. **Fisker High School Students' Favorite Movies**

Foreign	12
Cartoons	18
Romance	24
Adventure	36
Science fiction	45

2. **Sports Participation at Fisker High School**

Tennis	16
Ice hockey	22
Soccer	36
Softball	40
Lacrosse	48

3. **Fisker High School Clubs**

Chess Club	15
Radio Club	27
French Club	10
Rocketry Club	31
Art Club	26

4. **Optional Courses at Fisker High School**

Archery	11
Typing	38
Drivers Education	22
Accounting	16
Cooking	35

Use with pages 382–383.

PRACTICE: Making a Table to Find a Pattern

Make a table and solve.

4. Rachael lifts weights. She is building up her strength by following a training schedule. During the first week, she used 12-lb weights. During the second week, she used 24-lb weights. During the third week, she used 30-lb weights. How heavy are the weights she used on the fifth week?

5. The school store stocks poster board. They stock 5 sheets of size A, 15 sheets of size B, 35 sheets of size C, and 75 sheets of size D. The sizes go up to size F. How many sheets of poster board does the store stock?

6. The manager is buying lined paper for the store. A package of 5 pads costs $4.75. A package of 10 pads costs $9.45. A package of 50 pads costs $47.00. How much will a package of 100 pads cost?

7. The manager is building bookcases for the store. The second shelf is 24 inches higher than the first shelf, the third shelf is 21 inches higher than the second, and the fourth shelf is $18\frac{3}{8}$ inches higher than the third. How much higher is the fifth shelf than the fourth?

8. Tony is getting ready for a swim meet. He swam a total of 7 laps after one day, 14 laps after two days, 22 laps after three days, and 31 laps after four days. How many laps will he have swum after one week?

Use with pages 384–385.

PRACTICE: Broken-Line Graph

Use the information in each table to make a broken-line graph.

1.

POPULATION OF THE UNITED STATES

Year	Population	Year	Population
1880	50,000,000	1940	132,000,000
1890	63,000,000	1950	151,000,000
1900	76,000,000	1960	179,000,000
1910	92,000,000	1970	203,000,000
1920	106,000,000	1980	221,000,000
1930	123,000,000		

2.

PERCENT OF FREESTONE COUNTY POPULATION LIVING IN RURAL AREAS

Year	Percent (%)	Year	Percent (%)
1900	74	1940	40
1910	60	1950	31
1920	50	1960	29
1930	44	1970	26

3.

POPULATION OF SMALLVILLE

Year	Population	Year	Population
1960	8,525	1975	10,250
1965	9,250	1980	10,575
1970	9,550	1985	11,025

PRACTICE Problem-Solving Practice

Write the letter of the correct formula to help solve each problem.

1. Mrs. Mack made a circular tablecloth for her new dining-room table. The diameter of the tablecloth is 60 inches. How much material is in the tablecloth?

   a. $A = \pi d$ b. $A = lw$ c. $A = \pi r^2$

2. Frank's father built a sandbox that measures 1.2 meters by 1.8 meters by 0.2 meters. How many cubic meters of sand are needed to fill the box?

   a. $V = lwh$ b. $V = 2l + 2w$ c. $V = \pi r^2 h$

Write the letter of the correct proportion.

3. Frances wants to make French Onion soup for her dinner party. Her recipe for 4 people calls for 2 cups of cheese and $1\frac{1}{2}$ cups of onions. She wants to make the recipe for 6 people. How much cheese does she need?

   a. $\frac{4}{6} = \frac{n}{2}$ b. $\frac{2}{4} = \frac{n}{6}$ c. $\frac{2}{6} = \frac{n}{4}$

Make a plan to solve each problem.

4. Mr. Pandi spent $52 to plant a garden. He sold vegetables from the garden and earned $540. He spent 24% of his profits on seeds for next year. How much did he spend on next year's seeds?

   Step 1: _____

   Step 2: _____

5. Mr. Pandi bought 108 tulip bulbs. He paid $5.21 for a dozen bulbs. Each pot that he sold for $4.65 contained 3 tulips. How much profit did he make when he sold all the tulips?

   Step 1: _____

   Step 2: _____

   Step 3: _____

Use with pages 388–389.

PRACTICE Problem-Solving Practice

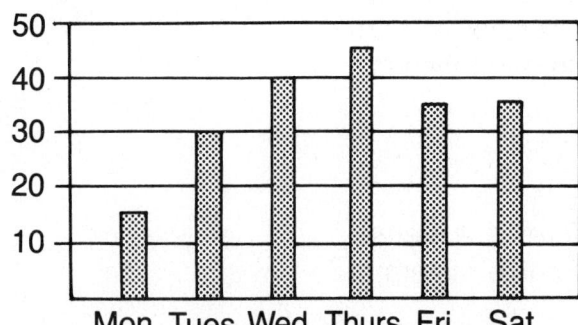

NUMBER OF FISH SOLD PER DAY AT WEST MALL PET STORE LAST WEEK

Use the bar graph to solve each problem.

6. What percent of the fish sold last week at the West Mall Pet Store was sold on Wednesday? _____

7. What percent of the fish sold last week at the West Mall Pet Store was sold in the first half of the week? _____

Solve.

8. Andrew has 5 boards. Each board is 4 feet long. He uses 3 feet of board to make a bird feeder. He wants to make 6 bird feeders. Does he have enough boards?

_____

9. Andrew figures that it took him $3\frac{1}{4}$ hours to make the first bird feeder and $2\frac{1}{2}$ hours to make each of the other 5. How many hours did it take Andrew to make all 6 bird feeders?

_____

10. Andrew sold each bird feeder for $2.75. His expenses totaled $6.72. If his expenses remain the same, how many more bird feeders can he make by using the profit on this sale to cover expenses? _____

11. Andrew figures that each bird feeder will hold 540 cubic inches of bird seed. The length of the bird feeder is 18 inches and the width is 10 inches. How high is the feeder?

_____

Use with pages 388–389.

PRACTICE: Check That the Solution Answers the Question

Which item answers the question? Write the letter of the correct answer.

1. During the month of June, West-Side Car Wash washed 538 cars. During the month of July, workers washed 627 cars. How many more cars did they wash in July than in June?
   a. More cars were washed in July.
   b. In July, they washed 627 cars.
   c. In July, they washed 89 more cars.

2. West-Side Car Wash charges $5.00 for a complete car wash. During August, workers washed 589 cars. Did West-Side receive more than $2,500 in August?
   a. West-Side received $2,945 in August.
   b. Yes, West-Side received more than $2,500 in August.
   c. No, West-Side received less than $2,500 in August.

Solve.

3. West-Side Car Wash has 12 employees. Of that number, $\frac{2}{3}$ work part time. How many employees work full time?

4. Mary Sue is the cashier for the car wash. She has been working at that job for $3\frac{1}{2}$ years. Every six months, Mary Sue gets one week of paid vacation. Has Mary Sue earned more than 2 months of vacation so far?

5. West-Side Car Wash replaces the brushes on its scrubbers every 3 months. West-Side has been in business 6 years. Has the company replaced the brushes more than 20 times?

6. West-Side Car Wash spends 12% of its monthly budget for TV advertisements and 8% for newspaper ads. If the monthly budget is $4,380, how much does West-Side spend on advertisements per month?

7. Frank Thomas has been the manager of the car wash for 5 years. Each year, he receives a $525 bonus. Has Mr. Thomas received $2,500 in bonuses yet?

8. West-Side Car Wash washed 4,708 cars and vans last year. The company does not wash trucks. The number of vans washed was about $\frac{1}{8}$ of the total number of vehicles washed. How many vans did West-Side wash last year?

Use with pages 390–391.

PRACTICE: Probability and Expectation

Pick a piece of paper out of the hat.

1. What are the possible outcomes? _____

2. Are they equally likely outcomes? _____

3. What is the probability that Diandre's name is selected? _____

4. What is the probability that a boy's name is selected? _____

5. What is the probability that Isabel's name is selected? _____

Suppose you spin the arrow once. Write a fraction for each probability.

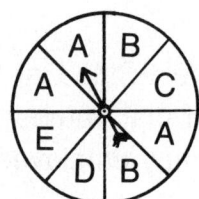

6. A _____ 　　7. B _____ 　　8. C _____

9. A or D _____ 　10. A or B _____ 　11. B or C _____

12. a letter from the word BEAD _____ 　　13. Which letter is most likely? _____

14. Which letter is least likely? _____

15. What is the probability of not getting an E? _____

Jane works for the school entertainment committee. She has to organize a party for Parents' Day. She is asked to choose at random one of these entertainers: George Fischer, musician; Fanny Andrews, dancer; Albert Jones, comedian; Susan King, musician. Write the probability of each.

16. She will pick a musician. _____ 　　17. She will pick Susan King. _____

18. She will pick a male entertainer. _____

19. She will pick neither a musician nor a dancer. _____

20. She will pick a magician. _____

PRACTICE: Independent Events

Solve.

1. List the 2-digit numbers that can be made from 5, 6, and 7.

   _____

2. George has tickets for track-and-field events. He has to choose which events to attend since some are scheduled at the same time. Draw a tree diagram that shows the combinations of events he can attend.

   > 10:00—mile run, hurdles
   >
   > 11:00—long jump, 100-m run, high jump
   >
   > 12:00—shot put, pole vault

3. To choose which member of the team will compete first, Yolanda, George, Cara, and Francisco decide to have their coach pull one of their names out of a jar. The coach picks a name and places it back in the jar. The team decides to let her use the same method to choose who will give a speech at the awards ceremony after the games. What is the probability that Yolanda's name will be picked both times?

   _____

4. There are 7 gymnasts on the team: Lee, Aido, Sue, Gene, Lore, Jaime, and Jon. Each can enter any one of 3 events: balance beam, floor exercise, or vaulting. Write the number of possible combinations.

   _____

5. The members of the cycling team are Britt, Wade, Theo, and Marla. Each can train in a city, mountain, valley, or seaside location. Write the number of possible combinations.

   _____

Use with pages 394–395.

PRACTICE — Integers

Write an integer to describe each situation.

1. a $10 profit _____

2. 6° below zero _____

3. 100 feet below sea level _____

4. a bank deposit of $50 _____

5. a $15 loss _____

6. 15° above zero _____

7. 300 feet above sea level _____

8. a loss of 8 points _____

9. a rise of 3° in temperature _____

10. a descent of 73 feet below sea level _____

Write the opposite situation and the opposite integer.

11. a drop of 10° in temperature
_____

12. $500 loss
_____

13. a climb of 100 yards
_____

14. $80 bank withdrawal
_____

Solve.

15. A plane flies at an altitude of 10,000 feet. What integer would show its descent to the ground?
_____

16. A football team loses 20 yards on a play. What integer would show the loss?
_____

PRACTICE Comparing Integers

Write >, <, or = for ○.

1. ⁻4 ○ ⁻5
2. ⁻1 ○ ⁺1
3. ⁻18 ○ ⁻19
4. ⁻11 ○ ⁺12
5. ⁺6 ○ ⁻7
6. ⁺8 ○ ⁻9
7. ⁺1 ○ ⁻1
8. ⁻8 ○ ⁺7
9. ⁺9 ○ 0
10. ⁺17 ○ ⁺18
11. 0 ○ ⁻6
12. ⁻11 ○ ⁺7
13. ⁺4 ○ ⁻3
14. ⁺10 ○ ⁺10
15. ⁺3 ○ ⁻3
16. ⁺5 ○ ⁻11
17. ⁺8 ○ ⁺7
18. ⁻14 ○ ⁻11
19. ⁺15 ○ ⁻15
20. ⁻8 ○ ⁺8

Write in order from the least to the greatest.

21. ⁻27, ⁻29, ⁻20
22. ⁻5, ⁻10, ⁻15
23. ⁺3, ⁺6, ⁺9

24. ⁺9, ⁻10, ⁺1
25. ⁺17, ⁻10, 0
26. ⁺1, 0, ⁻1

Write in order from the greatest to the least.

27. ⁺50, ⁻25, ⁺25
28. ⁻4, ⁻8, ⁻12
29. ⁻36, ⁺46, ⁺6

30. ⁺7, 0, ⁺3
31. ⁺7, ⁻7, ⁺10
32. ⁻13, ⁻6, ⁻1

33. Five families live in Pine Valley. The elevation of their homes, above or below sea level, are shown. List the families in order from the greatest to the least elevation of their homes.

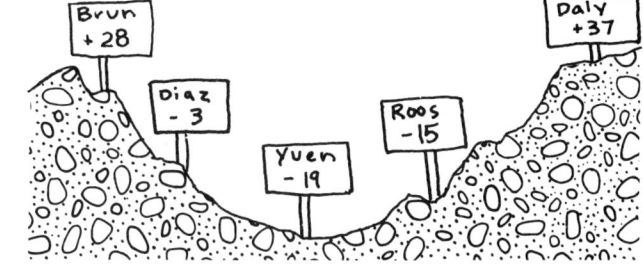

Brun +28
Diaz -3
Yuen -19
Roos -15
Daly +37

Use with pages 406–407.

PRACTICE: Adding Integers: Like Signs

Add.

1. ⁻5 + ⁻9 = _____
2. ⁻8 + ⁻13 = _____
3. ⁻6 + ⁻4 = _____
4. ⁻20 + ⁻2 = _____
5. ⁻16 + ⁻2 = _____
6. ⁻18 + ⁻18 = _____
7. ⁻5 + ⁻15 = _____
8. ⁻19 + ⁻1 = _____
9. ⁻1 + ⁻10 = _____
10. ⁺3 + ⁺6 = _____
11. ⁺9 + ⁺11 = _____
12. ⁺6 + ⁺7 = _____
13. ⁺2 + ⁺16 = _____
14. ⁺1 + ⁺21 = _____
15. ⁺4 + ⁺11 = _____
16. ⁻10 + ⁻4 = _____
17. ⁻18 + ⁻7 = _____
18. ⁻3 + ⁻22 = _____
19. ⁺17 + ⁺2 = _____
20. ⁺19 + ⁺7 = _____
21. ⁻6 + ⁻5 = _____
22. ⁺1 + ⁺3 = _____
23. ⁻7 + ⁻7 = _____
24. ⁻2 + ⁻1 = _____
25. ⁻12 + ⁻9 = _____
26. ⁻5 + ⁻20 = _____

Solve.

27. When Rick began to ice-skate, the temperature was 11°F. When he stopped, it had risen 9°. What was the temperature when he stopped?

_____

28. A football team lost 7 yards on one play. On the next play, the team lost 15 yards. After the second play, how far were they from their starting point?

_____

PRACTICE   Adding Integers: Unlike Signs

Add.

1. $^+2 + {^-13} =$ _____
2. $^+1 + {^-7} =$ _____
3. $^-20 + {^+5} =$ _____
4. $^-8 + {^+4} =$ _____
5. $^-16 + {^+18} =$ _____
6. $^-10 + {^+20} =$ _____
7. $^+15 + {^-20} =$ _____
8. $^+11 + {^-2} =$ _____
9. $^-8 + {^+12} =$ _____
10. $^-6 + {^+10} =$ _____
11. $^-13 + {^+3} =$ _____
12. $^+7 + {^-5} =$ _____
13. $^-5 + {^+30} =$ _____
14. $^+18 + {^-9} =$ _____
15. $^+8 + {^-2} =$ _____
16. $^+10 + {^-14} =$ _____
17. $^-15 + {^+19} =$ _____
18. $^-13 + {^+5} =$ _____
19. $^-1 + {^+6} =$ _____
20. $^-9 + {^+11} =$ _____
21. $^+7 + {^-7} =$ _____
22. $^+30 + {^-23} =$ _____
23. $^-3 + {^+11} =$ _____
24. $^-14 + {^+9} =$ _____
25. $^+5 + {^-15} =$ _____
26. $^+13 + {^-12} =$ _____
27. $0 + {^-12} =$ _____
28. $^-19 + {^+3} =$ _____
29. $^-16 + {^+16} =$ _____
30. $^+22 + {^-13} =$ _____

Solve.

31. At noon on a Monday in May, the temperature was 53°F. At sunset, the temperature was 11° lower. What was the temperature at sunset?

_____

32. A diver descends 72 ft from the ocean surface. After collecting some coral, she ascends 35 ft to observe a school of fish. How far below the surface is she then?

_____

Use with pages 410–411.

PRACTICE: Subtracting Integers

Subtract.

1. $^-4 - {^-7} =$ _____
2. $^+7 - {^-8} =$ _____
3. $^+2 - {^+8} =$ _____
4. $^-12 - {^+15} =$ _____
5. $^-6 - {^-1} =$ _____
6. $^+10 - {^-9} =$ _____
7. $^+12 - {^+13} =$ _____
8. $^-11 - {^+20} =$ _____
9. $^-9 - {^-11} =$ _____
10. $^-4 - {^+2} =$ _____
11. $^+5 - {^+14} =$ _____
12. $^-9 - {^+1} =$ _____
13. $^-3 - {^+1} =$ _____
14. $^+6 - {^-3} =$ _____
15. $^+1 - {^+19} =$ _____
16. $^-4 - {^-5} =$ _____
17. $^-8 - {^+6} =$ _____
18. $^+16 - {^+6} =$ _____
19. $^+7 - {^-2} =$ _____
20. $^-24 - {^-12} =$ _____
21. $^+2 - {^-18} =$ _____
22. $^+15 - {^+9} =$ _____
23. $^-8 - {^+3} =$ _____
24. $^+5 - {^-17} =$ _____
25. $^+14 - {^-14} =$ _____
26. $^+20 - {^-3} =$ _____
27. $^-16 - {^+11} =$ _____
28. $^-13 - {^+19} =$ _____
29. $^-10 - {^-2} =$ _____
30. $^-6 - {^-6} =$ _____

Birth Dates of Famous Thinkers in History

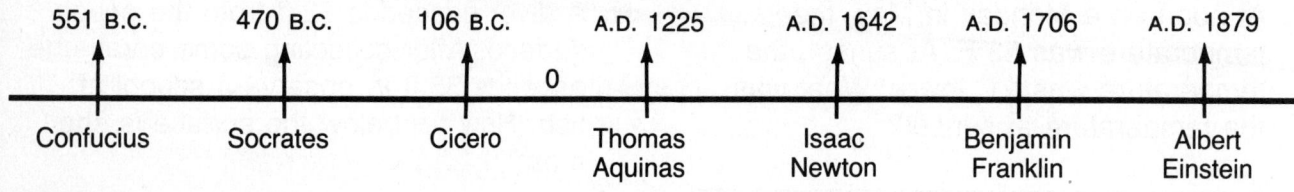

| 551 B.C. | 470 B.C. | 106 B.C. | A.D. 1225 | A.D. 1642 | A.D. 1706 | A.D. 1879 |
| Confucius | Socrates | Cicero | Thomas Aquinas | Isaac Newton | Benjamin Franklin | Albert Einstein |

Use the dates shown on the time line. How many years passed between the birth dates of

31. Confucius and Cicero? _____
32. Socrates and Isaac Newton? _____
33. Thomas Aquinas and Benjamin Franklin? _____
34. Cicero and Albert Einstein? _____
35. Confucius and Isaac Newton? _____

Use with pages 412–413.

PRACTICE: Using a Line Graph

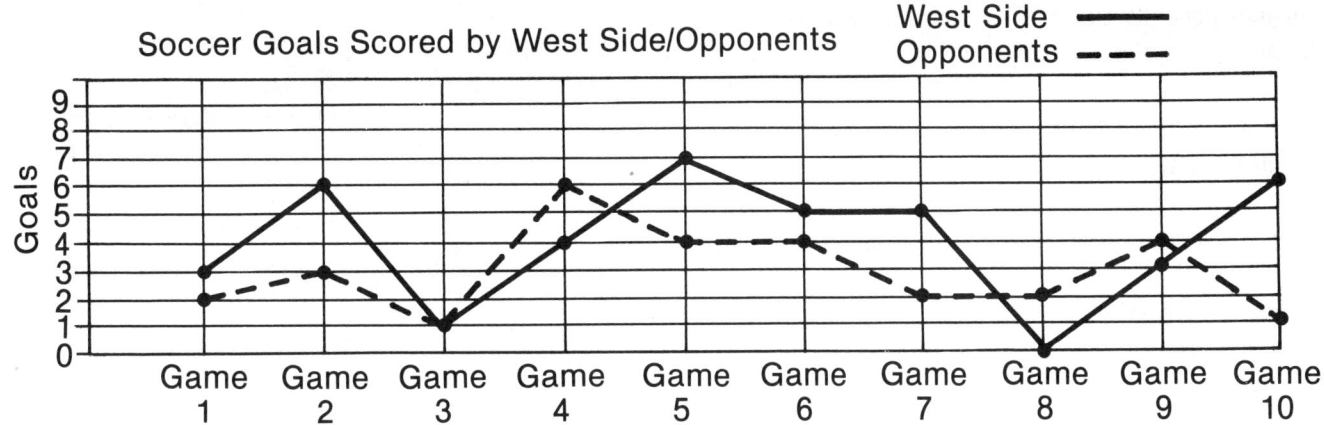

Use the line graph to answer these questions. Write *yes* or *no*.

1. Did West Side have a winning season? _____

2. Did West Side win more games during the first half of the season rather than the second half? _____

Use the graph to solve each problem.

3. How many games did West Side win? _____

4. How many games ended in a tie? _____

5. How many goals did West Side score during the season? _____

6. How many more goals did West Side score than its opponents? _____

7. What was the average number of goals per game scored by West Side? _____

8. In which game was the least number of goals scored by Westside? _____

9. Which games were won by one point? _____

10. By how many points did West Side win in its highest-scoring game? _____

11. By how many points did West Side win in the game in which the team had the greatest lead? _____

12. Which game do you think was the most exciting to watch? Why? _____

Use with pages 414–415.

167

PRACTICE: Graphing Ordered Pairs

The Botanical Garden is about to open several new exhibits. The locations are shown on the grid.

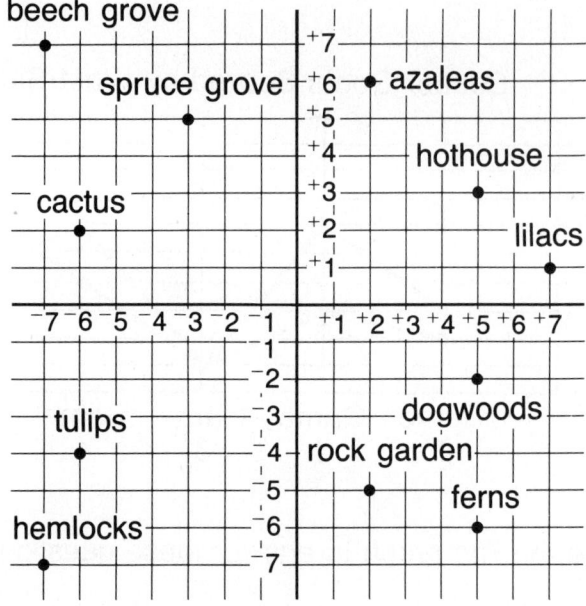

Write an ordered pair to describe the location of the

1. dogwoods. _____
2. spruce grove. _____
3. rock garden. _____
4. azaleas. _____
5. hemlocks. _____

Write the name of the exhibit identified by the ordered pair.

6. $(^+7, ^+1)$ _____
7. $(^-7, ^+7)$ _____
8. $(^-6, ^-4)$ _____
9. $(^+5, ^-6)$ _____
10. $(^+5, ^+3)$ _____
11. $(^-6, ^+2)$ _____

Use graph paper to make a grid like the one on the right.

Graph these points. Connect them in order to form a figure.

$(^+1, ^+5), (^+1, ^+2), (^+2, ^+2), (^+2, ^-2), (^+3, ^-2),$
$(^+3, ^-6), (^-3, ^-6), (^-3, ^-2), (^-2, ^-2), (^-2, ^+2),$
$(^-1, ^+2), (^-1, ^+5)$

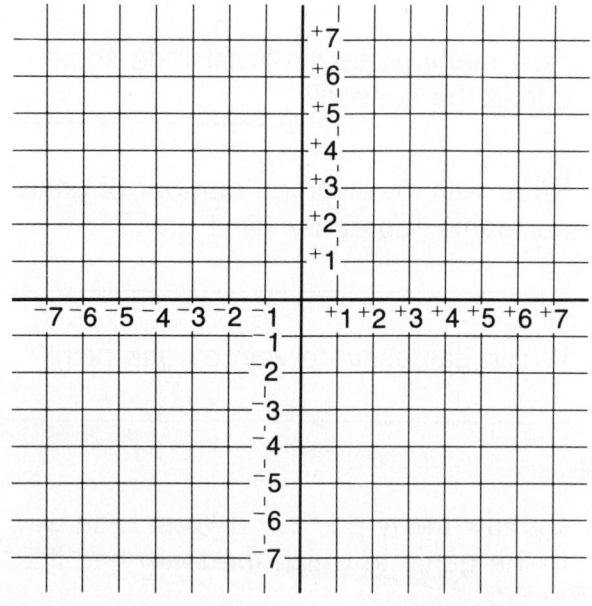

168   Use with pages 416–417.

PRACTICE Problem Solving Practice

Make a table and solve.

1. A new electronics firm is opening in Westview. The firm, Trontech, needs to hire employees. During the first week, they plan to hire 10 skilled workers and 15 unskilled workers. During the second week, they plan to hire 20 skilled workers and 12 unskilled workers. During the third week, they plan to hire 30 skilled workers and 9 unskilled workers. Trontech will be hiring for 5 weeks. If they hire skilled and unskilled workers according to this pattern, how many employees in all will Trontech hire?

To solve, make a guess, and check your answer.

2. Trontech is located on Industrial Parkway. The address number is a 3-digit number. It is divisible by 11. The first and last digits are the same, and the sum of the digits is 12. What is the address number?

Solve. Draw a picture if needed.

3. Mary Santiago attends a series of meetings held in offices along one hall. After she leaves the first office, she goes to the office 3 doors down the hall. After that meeting, she walks to the office that is 2 doors up the hall. Then she goes to the office 4 doors down. Next she walks to the office 1 door up, and finally to the last office, 2 doors down. How many offices are there along the hall?

Make a list and solve.

4. Frank Perry is an engineer at Trontech. He is experimenting with combinations of electronic boards. He makes boards 16 cm long by using combinations of boards lengths 10 cm, 5 cm, 3 cm, and 1 cm. He has three 10-cm boards, one 5-cm board, three 3-cm boards, and four 1-cm boards. How many 16-cm boards can Frank put together?

Use with pages 418–419.

PRACTICE: Problem-Solving Practice

Solve.

5. Ed Jones places one memory chip and one processor chip on a special board. He has 4 kinds of memory chips and 3 kinds of processor chips. How many special boards can he put together? _____

6. Each electronic part made by Trontech has an identification number. This is a 3-digit number. One of the identification numbers has no divisors other than 1. The last digit is 1. The sum of the first and second digits is 3 times the last digit. The second digit subtracted from the first digit equals the last digit. What is the identification number? _____

7. Some electronic experiments are very expensive. On one experiment, Trontech spent $600 for processor chips. The company spent $\frac{2}{3}$ of the remaining money on memory chips and $225 on boards. The company spent $75 for miscellaneous small parts. How much money did Trontech spend on the experiment? _____

8. Lin Chen described different kinds of computer boards to a group of visitors. The boards were placed in a long row. She talked about the first board in the row, then skipped ahead 4 boards, and talked about the next board. She then skipped back 2 boards and talked about that board. Then, she skipped over 4 boards and talked about the last one. How many boards were placed in the row?

_____

9. Trontech has 200 employees. The yearly expense for salaries amounts to $5,000,000. They expect expenses for 300 salaries to be $9,000,000. They expect expenses for 400 salaries to be $14,000,000. What will be the expenses for salaries if Trontech employs 500 workers?

_____

TO THE TEACHER:

RETEACH

This section provides additional instruction on material covered in the pupil's edition. Where appropriate, the emphasis of the instruction is on remediation of common errors. Each page is keyed to the appropriate lesson, but could be used as remediation, extra practice, or review at any time after the lesson.

RETEACH: Numbers to Hundred Thousands

A popular TV program sold advertising time at a rate of four hundred fifty thousand dollars for 30 seconds of air time. Write the number for this dollar amount.

Remember

Make a place-value chart.

Hundred thousands	Ten thousands	Thousands	Hundreds	Ten	Ones
4	5	0	0	0	0

Four hundred fifty thousand means four hundred thousand and fifty thousand. Write a 4 in the hundred thousands place and a 5 in the ten thousands place. Write a 0 in each other place.

The number amount is $450,000.

Write the number in standard form.

1. five hundred sixty-eight thousand, four hundred _____

2. one hundred fifty thousand _____

3. three hundred fifteen thousand, seven hundred forty-six
_____

4. 700,000 + 40,000 + 9,000 + 300 + 80 + 5 _____

5. 100,000 + 20,000 + 3,000 + 400 + 50 + 6 _____

6. 200,000 + 50,000 + 4,000 + 700 + 60 + 8 _____

7. 500,000 + 9,000 + 20 + 2 _____

8. Garry Glamour, the famous actor, was paid two hundred fifty-four thousand, six hundred ninety-two dollars for each episode of his prime-time hit *South Passaic*. Write his salary per episode in numbers. _____

9. Pearl Purehart has been a star of the daytime rural drama *All My Chickens* for 21 years, and has *never* missed an episode. She has appeared in one hundred eighty-seven thousand, two hundred one scenes in all. Write the number for this amount. _____

Use with pages 2–3.

RETEACH — Numbers to Hundred Billions

The planet Pluto is seven billion, three hundred seventy-five million kilometers from the sun. Write this number in standard form.

Remember

Make a place-value chart.

Billions			Millions			Thousands			Ones		
hundred billions	ten billions	billions	hundred millions	ten millions	millions	hundred thousands	ten thousands	thousands	hundreds	tens	ones
		7	3	7	5	0	0	0	0	0	0

Seven billion, three hundred seventy-five million means seven billions and three hundred seventy-five millions. Write a 7 in the billions place. Write a 3 in the hundred millions place, a 7 in the ten millions place, and a 5 in the millions place.

The number is 7,375,000,000.

Write the number in standard form.

1. six hundred forty-three billion, twenty-one million, eight hundred two thousand, five hundred sixty-eight _____

2. twenty billion, six hundred million, eight _____

3. one billion, five hundred thousand _____

Write the place value of the 3 in each number.

4. 361,475,682,610

5. 931,645,000,000

6. 828,643,712,000

7. 100,102,030,000

8. 163,000

9. 820,345,111

2 Use with pages 4–5.

RETEACH: Comparing and Ordering Numbers

The table shows the diameter of the planets in our solar system. Which planet is smaller, Mercury or Venus?

Planet	Diameter
Mercury	4,880 km
Venus	12,104 km
Earth	12,756 km
Mars	6,787 km
Jupiter	142,800 km
Saturn	120,000 km
Uranus	51,800 km
Neptune	49,500 km
Pluto	6,000 km

Remember

To compare numbers, align them on a place-value chart. Then compare the digits in each place.

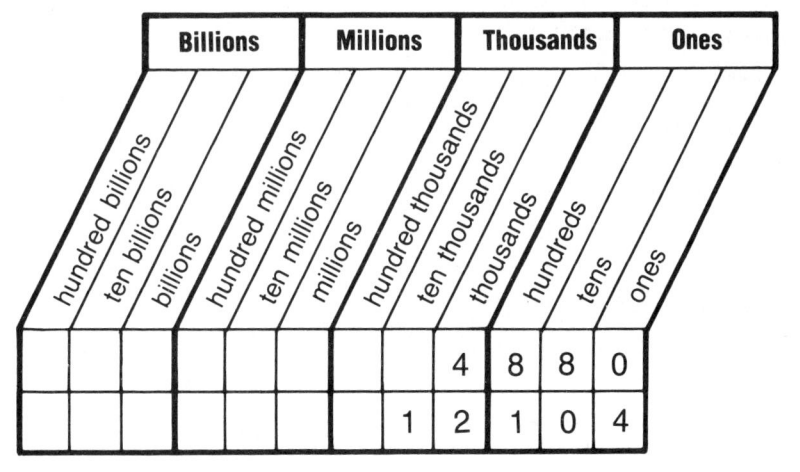

Mercury is smaller.

Compare. Write >, <, or =.

1. 4,000 _____ 898

2. 531,890 _____ 67,835

3. 3,281,093 _____ 3,427,586

4. 11,100 _____ 10,100

5. 4,536 _____ 5,436

6. 9,187 _____ 87,426

7. 138,645 _____ 138,545

8. 837,426 _____ 874,426

9. List the planets in the table above in order from the largest to the smallest.

_____

_____

Use with pages 6–7.

RETEACH: Addition and Subtraction Facts and Properties

Randy and Elaine attended the county fair and decided to play the beanbag toss. On his first try, Randy won 5 balloons. On his second try, he won 3 more. Elaine tried next and won 3 balloons on her first try. On her second try, she won 5 more. Who won more balloons, Randy or Elaine?

> **Remember**
>
> When numbers are added together, it doesn't matter in what order they are added. This is known as the **Commutative Property of Addition.**
>
> You can reverse the numbers without changing the sum.
>
> $$5 + 3 = 8 \quad 3 + 5 = 8$$
>
> When you add more than two numbers, you can use another property of addition called the **Associative Property.** Look at the example below.
>
> $$(2 + 1) + 6 = 9 \quad 2 + (1 + 6) = 9$$
>
> No matter which two numbers are added first, the sum is always the same.

Randy and Elaine won the same number of balloons.

Find the missing number.

1. $3 + 4 = 4 + \underline{}$
2. $6 + 2 = 2 + \underline{}$
3. $\underline{} + 8 = 8 + 2$
4. $9 + \underline{} = 3 + 9$
5. $(3 + 4) + 5 = 3 + (4 + \underline{})$
6. $(7 + 2) + 6 = 7 + (2 + \underline{})$
7. $(8 + 1) + 4 = \underline{} + (1 + 4)$
8. $(\underline{} + 3) + 7 = 5 + (3 + 7)$
9. $(11 + \underline{}) + 12 = 11 + (13 + 12)$
10. $(9 + 6) + \underline{} = 9 + (6 + 4)$

Solve.

11. Randy and Elaine stayed at the fair all day. During the day, they both rode on the same number of rides. In the morning, Randy went on 5 rides, took 2 more rides in the afternoon, and went for 4 rides in the evening. Elaine went on 4 rides in the morning and on 5 more rides in the afternoon. How many rides did she take in the evening?

    _____

Use with pages 10–11.

RETEACH: Estimating Sums and Differences—Front-End

Tom has $15 to spend on phonograph records. Will that be enough to pay this bill?

Music City
Records: $6.95
3.49
5.69
Total _____

Remember

Adjust by grouping cents to dollars.

$6.95 → about $1
3.49 ⎫
5.69 ⎭ → about $1

Dollars: 6 + 3 + 5 = $14
Adjustment: about $2
Estimate: $14 + $2 = $16

Tom does not have enough to pay this bill.

Estimate the sum or difference. Choose the best answer.

1. $2.35 + $1.69 + $3.45   **a.** under $7   **b.** over $7

2. $8.89 + $6.29 + $2.65   **a.** under $20   **b.** over $20

3. 4,125 + 3,680 + 1,010   **a.** under 8,000   **b.** over 8,000

4. 24,250 + 13,062 + 65,947   **a.** under 100,000   **b.** over 100,000

5. 6,238 + 507 + 3,615   **a.** under 11,000   **b.** over 11,000

6. $428.39 − $259.75   **a.** under $200   **b.** over $200

7. $783.60 − $335.19   **a.** under $400   **b.** over $400

8. 3,875 − 580   **a.** under 3,000   **b.** over 3,000

Estimate.

9.   2,117
     5,251
       103
   + 4,368

10. $3.28
      2.49
      2.85
   + 4.53

11.   15,680
   − 9,725

12. $823.19
   − 549.10

Use with pages 12–13.

RETEACH: Rounding and Estimating Sums & Differences

Round 36,228 to the nearest thousand.

Remember

Be sure to round to the correct place value.

Find the thousands place.	Look at the digit in the hundreds place.	If the digit is • 5 or greater, round up; • Less than 5, round down.
3⑥,228	36,②28	2 < 5 Round down.

Round $637.49 to the nearest ten dollars.

Find the tens place.	Look at the digit in the ones place.	7 > 5 Round up.
$6③7.49	$63⑦.49	

The number 36,228 rounded to the nearest thousand is 36,000. The number $637.49 rounded to the nearest ten dollars is $640.00.

Round to the nearest thousand.

1. 5,329 _____
2. 126,710 _____
3. 59,507 _____

Round to the nearest million.

4. 28,512,704 _____
5. 123,706,217 _____
6. 8,325,495 _____

Round to the nearest dollar.

7. $23.62 _____
8. $419.80 _____
9. $65.50 _____

Estimate.

10. $428.39
    + 579.25

11. 53,012
    + 62,170

12. 6,528
    + 8,119

13. 74,234
    + 12,896

14. $832.40
    + 285.14

15. 7,610
    − 2,580

16. 358,339
    − 139,025

17. 51,023
    − 25,117

Use with pages 14–15.

RETEACH Adding Whole Numbers

Earl had 58 photographs of different countries from around the world. His Uncle Albert gave him 88 photographs from his collection. How many photographs does Earl have now?

Remember

Regroup before adding.

Add the ones.   Regroup.   Add the tens.

```
          1              1
   5 8       5 8         ⑤ 8
 + 8 8     + 8 8       + ⑧ 8
 ─────     ─────       ─────
     ⑥         ⑥       ①④ 6
```

Regroup 1 ten and 6 ones.

Earl has 146 photographs.

Add.

1. 22   2. 16   3. 537   4. 668   5. 231
  + 71   + 27   + 15   + 24   + 16

6. 946   7. 802   8. 901   9. 736   10. 234
  + 712   + 617   + 126   + 51   + 942

11. 3,331 12. 4,436   13. 5,012   14. 5,447   15. 1,943
  + 3,942   +   162   + 7,044   + 1,831   + 1,211

Solve.

16. Bert wanted to visit the highest and lowest points on Earth. He climbed to the top of Mt. Everest, 8,848 meters above sea level. He then travelled to the bottom of the Mariana Trench, 10,912 meters below sea level. What was the total distance of these two trips?

Use with pages 18–19.

RETEACH: Column Addition

Grandiose Airlines makes a transatlantic flight every day from New York to Paris. The plane can carry 258 people in the First Class section, 164 people in the Top Drawer section, and 108 people in the Extremely Deluxe section. How many people in all can the flight carry from New York to Paris?

Remember

Regroup before adding.

Add the ones. Regroup.	Add the tens. Regroup.	Add the hundreds.
2	1 2	1 2
258	258	258
164	164	164
+ 108	+ 108	+ 108
0	30	530

The flight from New York to Paris can carry 530 people in all.

Add.

1. 94
   77
   + 74

2. 963
   896
   + 578

3. 410
   204
   + 486

4. 619
   535
   + 913

5. $8.13
   2.02
   + 4.95

6. 18,021
   29,114
   + 4,078

7. 27,652
   32,836
   + 20,651

8. 728,714
   1,759
   + 36,117

9. 63,346
   1,752
   + 42,965

10. $184.48
    883.75
    + 28.47

Solve.

11. During his flight to Paris, Lionel Eatwell noted the number of calories in every course of his evening meal so he could stay on his diet. The Paté Deluxe had 820 calories, the Chicken Grandiose had 784 calories, and the Chi Chi Flambé had 999 calories. How many calories were there in Mr. Eatwell's meal?

RETEACH   Subtracting Whole Numbers

From 1925 to 1939, Lou Gehrig played 2,130 baseball games in a row. The closest to this record that any player has come is 1,307 games. By how many games is Lou Gehrig's record still ahead?

Remember

Regroup before subtracting.

Regroup ones.	Regroup tens.	Regroup hundreds.	Subtract.
2 10	2 10	1 11 2 10	1 11 2 10
2,1 3 0̸	2,1 3̸ 0̸	2,1̸ 3̸ 0̸	2,1̸ 3̸ 0̸
− 1,3 0 7	− 1,3 0 7	− 1,3 0 7	− 1,3 0 7
			8 2 3

Regroup as 11 hundreds, 2 tens, and 10 ones. Subtract.

Lou Gehrig's record is still ahead by 793 games.

Subtract.

1. 994
   − 472

2. 186
   − 164

3. 913
   − 426

4. 999
   − 251

5. $8.67
   − 1.33

6. 7,232
   − 6,587

7. 4,488
   − 2,163

8. 8,143
   − 1,555

9. 9,332
   − 7,446

10. $77.86
   − 61.41

Solve.

11. The longest home run on record measured 618 feet and was hit by "Dizzy" Carlyle in 1929. Babe Ruth's longest home run measured 587 feet and was hit in 1919. How much longer was Carlyle's home run?

12. Babe Ruth had a lifetime record of 714 home runs. In 1976, Hank Aaron retired from baseball with a lifetime record of 755 home runs. By how many home runs did Hank Aaron overtake Babe Ruth's record?

Use with pages 22–23.

RETEACH: Subtracting with Zeros

The area of Alaska is 591,000 square miles. The area of Texas is 266,807 square miles. How much larger is Alaska than Texas?

Remember

Regroup before subtracting.

Regroup the ten thousands.	Regroup the thousands.	Regroup the hundreds.	Regroup the ones.
8 11 5 9̸ 1 , 0 0 0 − 2 6 6 , 8 0 7	10 8 1̸1̸ 10 5 9̸ 1 , 0̸ 0 0 − 2 6 6 , 8 0 7	10 9 8 1̸1̸ 1̸0̸ 10 5 9̸ 1 , 0̸ 0̸ 0 − 2 6 6 , 8 0 7	10 9 9 8 1̸1̸ 1̸0̸1̸0̸ 10 5 9̸ 1 , 0̸ 0̸ 0̸ − 2 6 6 , 8 0 7 3 2 4 , 1 9 3

Subtract.

Alaska is 324,193 square miles larger than Texas.

Subtract.

1. 360
    − 233

2. 700
    − 698

3. 960
    − 915

4. 900
    − 381

5. 75,005
    − 67,109

6. 5,000
    − 1,643

7. 8,340
    − 3,111

8. 7,850
    − 1,721

9. 9,570
    − 6,431

10. 200,019
    − 126,120

Solve.

11. The area of the island of Hawaii is greater than 4,000 square miles. The area of Rhode Island is 1,210 square miles. How much larger is the island of Hawaii than Rhode Island?

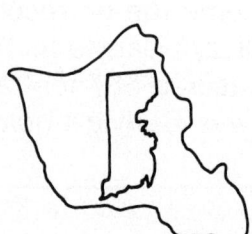

12. South Dakota has an area of 77,000 square miles. The area of North Dakota is 70,702 square miles. How much larger is South Dakota?

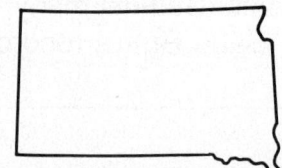 

Use with pages 24–25.

RETEACH   Equations

Karen and Tom work in a restaurant. Last night, they filled 30 relish trays. If Karen filled 16 relish trays, how many trays did Tom fill?

Remember

Think the problem through. There are 30 trays in all. Karen fills 16 of them. Tom filled the rest. Let n equal the number of trays Tom filled.

$$\text{Then } 30 - 16 = n.$$
$$\text{Subtract. } 30 - 16 = 14$$

Tom filled 14 relish trays.

Find the missing number.

1. $6 - \underline{} = 5$
2. $10 - \underline{} = 6$
3. $15 + \underline{} = 22$

4. $7 + 6 = \underline{}$
5. $\underline{} = 13 + 6$
6. $4 + 11 = \underline{}$

7. $13 - \underline{} = 7$
8. $\underline{} - 5 = 9$
9. $7 - 2 = \underline{}$

Solve.

10. One night at work, Karen served 17 slices of apple bread. She served 8 of them with a slice of cheddar cheese on the side. How many slices did she serve without cheese? _____

11. Tom needed 21 soupspoons for the buffet table. He already had 11 soupspoons. How many more soupspoons did he need? _____

12. At the end of the night, Karen and Tom divided their tips. They each took home 12 dollars in tips. How much money did they earn in tips altogether? _____

13. Karen and Tom served potato salad to a total of 14 customers. If Karen served 9 potato salads, how many did Tom serve? _____

Use with pages 28–29.

RETEACH Tenths and Hundredths

Kim and Lisa went for a 10-mile trail ride in the Theodore Roosevelt National Park. After riding for 7 miles, they stopped for a picnic. What part of the trail had they completed?

Remember
Compare the miles completed to the total miles.
$$\frac{\text{miles completed}}{\text{total miles}} = \frac{7}{10}$$

They had completed $\frac{7}{10}$ of the ride.

$\frac{7}{10}$ can be expressed in many ways.

| 7 of 10 parts shaded | In words: seven tenths | As a fraction: $\frac{7}{10}$ | As a decimal: Ones \| Tenths
0 . 7
decimal point |

Write as a decimal.

1.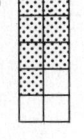
2.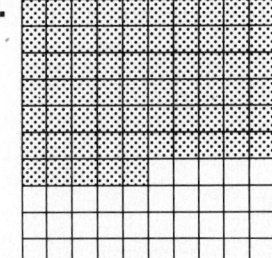
3.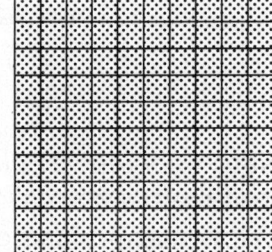
   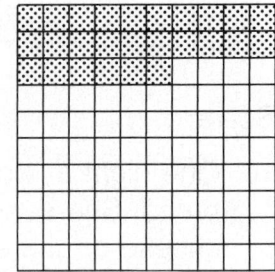

_____   _____   _____   _____

Copy and complete the chart. Write each number as a decimal.

	Hundreds	Tens	Ones	Tenths	Hundredths
4. $3\frac{7}{10}$					
5. $324\frac{72}{100}$					
6. $84\frac{3}{10}$					
7. $7\frac{49}{100}$					
8. $651\frac{7}{100}$					

RETEACH: Ten-Thousandths

A mechanic's shop lists the dimensions of car parts in decimals. One piston ring has a diameter of 3.6875 inches.

Write the word name for this number.

Remember

Read the word *and* after the ones place. Locate the decimal place value on the chart.

Tens	Ones	Tenths	Hundredths	Thousandths	Ten Thousandths
	3	6	8	7	5

Use the name of the place in which the last digit is found. Since 5 is in the ten-thousandths place, say five "ten-thousandths."

The word name is: three *and* six thousand eight hundred seventy-five ten-thousandths.

Give the value as a decimal of the underlined digit.

1. 7.<u>3</u>54 _____
2. 8.0<u>7</u>8 _____
3. 0.61<u>5</u> _____
4. 12.6<u>2</u>0 _____
5. 32.70<u>9</u> _____
6. 0.000<u>4</u> _____
7. 121.<u>3</u>70 _____
8. 9.4<u>8</u>7 _____

Write as a decimal.

9. 362 thousandths _____
10. 37 thousandths _____
11. 29 hundredths _____
12. 3 and 97 hundredths _____
13. 5,280 ten-thousandths _____
14. 6 and 12,975 hundred-thousandths _____

Solve.

15. The mechanic who repairs trucks needs oversized piston rings. She wants rings that are the three ten-thousandths size. Write the decimal for this. _____

Use with pages 40–41.

RETEACH: Comparing and Ordering Decimals

A sand sample contains 0.13 lb gravel, 0.0685 lb coarse sand, and 0.095 lb fine sand. Arrange these numbers in order from the greatest to the least.

Remember

Write zeros before comparing decimals so that the numbers are equivalent decimals.

> 0.13 ⟶ 0.1300
> 0.0685 ⟶ 0.0685
> 0.095 ⟶ 0.0950

Each number has the same number of digits after the decimal point. Line up the decimal points. Compare digits.

> 0.1300   Since 1 > 0, 0.1300 is the largest number.
> 0.0685
> 0.0950

Next, compare the digits in the hundredths place for the other two numbers.

> 0.0685   Since 9 > 6, 0.0950 > 0.0685.
> 0.0950

The numbers arranged from largest to smallest are:

      0.13   0.095   0.0685

Compare. Write >, <, or = for ◯.

1. 0.964 ◯ 9.35
2. 0.53 ◯ 0.523
3. 0.096 ◯ 0.069
4. 1.30 ◯ 1.3
5. 0.985 ◯ 0.958
6. 52.2 ◯ 52.9
7. 0.02 ◯ 0.002
8. 1.560 ◯ 1.56
9. 3.010 ◯ 3.100
10. 0.134 ◯ 0.156
11. 6.89 ◯ 7.99
12. 5.024 ◯ 5.025
13. 4.14 ◯ 4.14
14. 17.17 ◯ 17.71
15. 8.94 ◯ 8.91

Use with pages 44–45.

RETEACH — Rounding Decimals

Coins are measured in grains. A nickel has a mass of 73.1667 grains. Round this number to the nearest hundredth of a grain.

Remember

Look at the digit in the thousandths place to round to the nearest hundredth.

73.1667 ← thousandths place

Since the digit in the thousandths place is greater than 5, increase the digit in the hundredths place by 1.

73.1667 — Increase by 1. / greater than 5

Rounded to the nearest hundredth, 73.1667 becomes 73.17.

Round to the nearest whole number.

1. 4.35 _____
2. 24.42 _____
3. 91.71 _____
4. 29.10 _____
5. 38.5 _____
6. 4.31 _____
7. 93.17 _____
8. 13.60 _____
9. 34.06 _____
10. 78.9 _____

Round to the nearest tenth.

11. 24.31 _____
12. 1.49 _____
13. 9.46 _____
14. 35.26 _____
15. 5.45 _____
16. 0.75 _____
17. 0.28 _____
18. 5.37 _____
19. 34.33 _____
20. 0.74 _____

Round to the nearest hundredth.

21. 2.319 _____
22. 54.501 _____
23. 9.641 _____
24. 15.455 _____
25. 39.008 _____
26. 15.118 _____
27. 0.938 _____
28. 0.478 _____

Solve.

29. A U.S. dime has a mass of 38.5833 grains. Round this number to the nearest grain. Round this number to the nearest tenth of a grain. _____

Use with pages 46–47.

RETEACH: Estimating Decimal Sums and Differences

You can estimate decimal sums and differences by using either front-end estimation or rounding method.

Estimate 11.25 + 19.765

Remember

When using the front-end method, examine the decimal parts after adding the whole numbers.

Front End

Add the whole numbers.

11 + 19 = 30

Adjust by examining the decimal parts.

0.25 + 0.765 is around 1.
30 + 1 = 31

Rounding

Round each number to the nearest whole number. Then add.

11.25 → 11
19.765 → + 20
          31

The estimate is 31.

Estimate. Write > or < for ◯.

1. 2.87 + 3.68 ◯ 6
2. 10.52 + 6.86 ◯ 17
3. 0.825 + 0.315 + 1.103 ◯ 2
4. 17.36 + 15.92 ◯ 30
5. 2.019 + 0.065 + 3.275 ◯ 6
6. 5.051 + 0.929 + 0.857 ◯ 7
7. 21.29 + 30.05 + 5.12 ◯ 55
8. 6.24 + 2.03 + 7.09 ◯ 15
9. 18.235 − 8.627 ◯ 10
10. 23.11 − 22.83 ◯ 1
11. 9.41 − 6.85 ◯ 3
12. 43.628 − 21.139 ◯ 20
13. 50 − 25.21 ◯ 25
14. 86.751 − 50.209 ◯ 30
15. 95.637 − 92.815 ◯ 3
16. 73.4 − 23.8 ◯ 50
17. 30.101 − 15.989 ◯ 15
18. 20.7 − 11.3 ◯ 8
19. 45.88 − 20.88 ◯ 24
20. 64.64 + 10.10 ◯ 74
21. 22.34 + 5.79 ◯ 25
22. 13.04 + 7.23 ◯ 21

Use with pages 50–51.

RETEACH: Adding with Decimals

A quarter has a mass of 96.45 grains and a half-dollar has a mass of 192.9 grains. Find the combined mass of one quarter and one half-dollar.

Remember

Align the decimal points; then add. Regroup if necessary.

```
   1
  196.45
 + 192.9
  289.35  ← Place the decimal point.
```

The two coins have a mass of 289.35 grains.

Add.

1. 10.12
  + 62.23

2. 4.30
  + 40.573

3. 52.321
  + 15.063

4. 18.113
  + 45.341

5. 6.13
  + 8.19

6. 57.48
  + 2.473

7. 6.435
  + 2.9

8. $9.84
  + 5.39

9. 0.2
    0.3
  + 0.8

10. 7.00
     5.96
  + 8.25

11. 7.651
     4.921
  + 8.649

12. $4.27
     3.19
  + 8.37

Solve.

13. A U.S. penny has a mass of 48 grains. A nickel has a mass of 73.1667 grains, a dime 38.5833 grains, a quarter 96.45 grains, and a half-dollar 192.9 grains. Find the total mass of one of each of these coins.

Use with pages 52–53.

RETEACH — Subtracting with Decimals

A wooden block has a mass of 1.135 kilograms. After soaking in water for 8 days, the block has a mass of 1.927 kg. How much water was absorbed by the block?

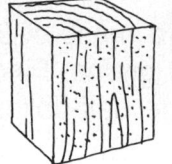

Remember

First subtract the thousandths, then the hundredths, the tenths, and the ones. Regroup if necessary.

Subtract the thousandths. Regroup.	Subtract the hundredths.	Subtract the tenths.	Subtract the ones.
8 12 1.9̸2̸7 − 1.1 3 5 ——— 2	8 12 1.9̸2̸7 − 1.1 3 5 ——— 9 2	8 12 1.9̸2̸7 − 1.1 3 5 ——— .7 9 2	8 12 1.9̸2̸7 − 1.1 3 5 ——— 0.7 9 2
		Place the decimal point.	

The water has a mass of 0.792 kg.

Subtract.

1. 8.46
   − 1.52

2. 19.050
   − 3.425

3. 0.8412
   − 0.8353

4. $3.19
   − 1.73

5. 8
   − 2.4

6. 18
   − 3.15

7. 12.3
   − 6.581

8. 9.3
   − 2.3752

9. 5.31 − 2.26 = _____

10. 0.895 − 0.27 = _____

11. 6.12 − 4.08 = _____

12. 21.543 − 19.678 = _____

13. 8.2 − 4.07 = _____

14. 13.32 − 7.8 = _____

15. 5.42 − 3.31 = _____

16. 8.91 − 3.5 = _____

RETEACH: Multiplication Facts and Properties

Zoom Airlines operates a small plane for short flights. It has seating on two levels. On the first level, there are 8 rows of seats with 2 seats in each row. On the second level, there are 11 rows of seats with 2 seats in each row. How many seats are there in the plane?

Remember

There are two ways to solve the problem. You can either add and then multiply, or multiply and then add.

$$(8 + 11) \times 2 = 38$$
$$(8 \times 2) + (11 \times 2) = 38$$

Add the total number of rows of seats and multiply by 2, or multiply the rows of seats in the first level by 2, multiply the rows in the second level by 2, and then add the two products.

There are 38 seats on the plane.

Find the missing number.

1. $5 \times (4 + 6) = (5 \times 4) + (5 \times \underline{\hspace{1cm}})$

2. $3 \times (2 + 5) = (3 \times 2) + (\underline{\hspace{1cm}} \times 5)$

3. $2 \times (4 + 3) = (2 \times \underline{\hspace{1cm}}) + (2 \times 3)$

4. $6 \times (3 + 5) = (\underline{\hspace{1cm}} \times 3) + (6 \times 5)$

5. $(6 \times 8) \times 7 = 6 \times (8 \times \underline{\hspace{1cm}})$

6. $(4 \times 3) \times 2 = 4 \times (\underline{\hspace{1cm}} \times 2)$

7. $5 \times 7 = \underline{\hspace{1cm}} \times 5$

8. $\underline{\hspace{1cm}} \times 3 = 3 \times 4$

Solve.

9. Zoom Airlines likes to make sure every passenger is comfortable during a flight. They keep 4 pillows in every closet on the plane in case anybody wants one. If there are 3 closets in the first section of the plane and 5 more closets in the second section of the plane, how many pillows are there in all? _____

Use with pages 70–71.

RETEACH: Multiplying with Multiples of 10; 100; 1,000

Each math-club member solved 40 problems in one hour. How many problems did all 9 math-club members solve in one hour?

Remember

Use expanded notation and the Associative Property to help keep the correct place value.

$$9 \times 40 = 9 \times (4 \times 10)$$
$$9 \times (4 \times 10) = (9 \times 4) \times 10$$
$$9 \times 40 = 360$$

The math club solved 360 problems in one hour.

Multiply.

1. 60 × 2
2. 70 × 5
3. 40 × 9
4. 50 × 3
5. 60 × 4

6. 600 × 30
7. 400 × 50
8. 300 × 80
9. 900 × 20
10. 600 × 60

11. 4,000 × 500
12. 6,000 × 200
13. 3,000 × 90
14. 3,000 × 700
15. 8,000 × 200

Use patterns to find the product.

16. 12 × 1 = _____
    12 × 10 = _____
    12 × 100 = _____
    12 × 1,000 = _____

17. 400 × 1 = _____
    400 × 10 = _____
    400 × 100 = _____
    400 × 1,000 = _____

Solve.

18. The math club was giving its annual banquet and reserved 20 tables for all the guests. If there are 10 people at each table, how many attended the banquet in all?
    _____

RETEACH — Estimating Products

If Louise gets paid $28 per week for babysitting, about how much will she earn in a year?

Remember

When one factor is rounded up and one factor is rounded down, no adjustment is necessary.

$$\$28 \longrightarrow \$30 \qquad 52 \longrightarrow 50$$

Multiply.

$$\$30.00 \times 50 = \$1,500.00$$

Adjust.
One factor rounded up and one down $\longrightarrow$ no adjustment necessary.

Louise will earn about $1,500.00 annually.

Mentally compute the product.

1. 20×70 _____
2. 500×400 _____
3. 38×100 _____
4. 60×50 _____
5. $200 \times 2,000$ _____
6. 30×510 _____

Write the letter of the best choice to estimate.

7. 42×98 **a.** 40×90 **b.** 40×100 **c.** 50×90 **d.** 50×100

8. 385×57 **a.** 60×30 **b.** 50×400 **c.** 60×400 **d.** $60 \times 4,000$

9. 519×63 **a.** 500×60 **b.** 500×70 **c.** 600×70 **d.** $5,000 \times 60$

10. 234×492 **a.** 200×50 **b.** 200×400 **c.** 200×500 **d.** $2,000 \times 500$

Estimate. Then write $>$ or $<$ to show how you would adjust the estimate.

11. 68×29
12. 32×23
13. 43×93
14. 86×79
15. 609×73
16. 584×32

17. 413×52
18. 729×19
19. 584×325
20. 115×426
21. 875×689
22. 968×423

Use with pages 74–75.

RETEACH: Multiplying by 1-Digit Factors

Gerry planted 4 packages of flower seeds in her garden. Each package contained 123 seeds. How many flower seeds were there in all?

Remember

Multiply the ones, then the tens, and finally the hundreds. Regroup when necessary.

Multiply the ones. Regroup the 1 ten.	Multiply the tens. Then add the 1 ten.	Multiply the hundreds.
1 123 $\times4$ $\overline{2}$	1 123 $\times4$ $\overline{92}$	123 $\times4$ $\overline{492}$

Multiply money the same way you multiply whole numbers. Then place the dollar sign and the cents point.

$$\begin{array}{r}1\\ \$1.42\\ \times6\\ \hline 2\end{array} \quad \begin{array}{r}2\,1\\ \$1.42\\ \times6\\ \hline 52\end{array} \quad \begin{array}{r}2\,1\\ \$1.42\\ \times6\\ \hline \$8.52\end{array}$$

There were 492 flower seeds in all.

Multiply.

1. 41 × 6
2. 41 × 4
3. 81 × 9
4. $0.47 × 2
5. $0.14 × 4

6. 85 × 6
7. 62 × 2
8. 54 × 6
9. $0.15 × 6
10. $0.52 × 4

11. 395 × 7
12. 813 × 9
13. 326 × 4
14. $3.96 × 7
15. $2.78 × 6

Solve.

16. Gerry likes tulips more than any other flower. She planted 3 rows of tulips in her garden. Each row had 25 plants in it. How many tulip plants did Gerry have in all?

RETEACH — Multiplying By 2-Digit Factors

During a recent storm at sea, some ocean waves traveled as fast as 36 miles per hour. How far did the waves travel in 24 hours?

Remember

Think of 36 as 30 + 6, or 3 tens + 6.

Multiply by ones.	Multiply by tens.	Add.
3 6 × 2 4 1 4 4	3 6 × 2 4 1 4 4 7 2 0	3 6 × 2 4 1 4 4 7 2 0 8 6 4

The waves traveled 864 miles.

Multiply.

1. 42 × 21
2. 21 × 31
3. 61 × 15
4. $0.41 × 12
5. $0.42 × 21

6. 82 × 25
7. 81 × 57
8. 44 × 60
9. $0.61 × 60
10. $0.82 × 92

11. 979 × 27
12. 713 × 31
13. 278 × 30
14. $2.01 × 73
15. $8.03 × 52

Solve.

16. The ocean current known as *the Gulf Stream* can travel 15 knots in 3 hours. At this rate, how far could the Gulf Stream travel in 30 hours? _____

17. A famous cruise ship can travel at a speed of 28 knots per hour. At this rate, how many knots could the ship travel in 12 hours? _____

Use with pages 80–81.

RETEACH: Multiplying by 3-Digit Factors

Southern Skies Airline operates 375 flights per day out of the airport in Atlanta. Each flight can carry 135 passengers. How many people can fly with this airline from Atlanta each day?

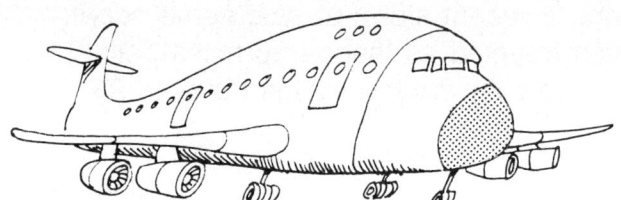

Remember

Multiply by the ones, the tens, and the hundreds. Be sure to watch place value. Regroup when necessary.

Multiply by the ones.	Multiply by the tens.	Multiply by the hundreds.	Add.
375 × 13**5** = 1,875	375 × 1**3**5 = 1,875; 11,250	375 × **1**35 = 1,875; 11,250; 37,500	375 × 135 = 1,875 + 11,250 + 37,500 = 50,625

Each day, 50,625 people can fly with the airline from Atlanta.

Multiply.

1. 832 × 546
2. 481 × 624
3. 971 × 505
4. $2.06 × 577
5. $8.11 × 203

6. 6,115 × 747
7. 9,237 × 614
8. 7,151 × 625
9. $58.38 × 643
10. $30.93 × 567

11. 9,307 × 241
12. 602 × 129
13. 633 × 201
14. $97.13 × 327
15. $82.63 × 694

RETEACH: Exponents and Squares

In one year, each of the 50 members of the Forestry Club planted 50 small trees. Use exponential form to express the number of trees planted in all.

Remember

Exponential form tells how many times a base number is used as a factor. In the equation 50 × 50, 50 has been used as a factor twice. Count the 50's. There are 2 of them; so, $50 \times 50 = 50^2$.

The club planted 50^2 trees in all.

Write in exponential form.

1. three to the second power _____
2. two to the fourth power _____
3. five to the third power _____
4. four to the fifth power _____
5. six to the second power _____
6. nine squared _____
7. eight squared _____
8. seven to the third power _____
9. $3 \times 3 \times 3 \times 3$ _____
10. $2 \times 2 \times 2$ _____
11. $4 \times 4 \times 4 \times 4 \times 4$ _____
12. $5 \times 5 \times 5$ _____
13. $9 \times 9 \times 9 \times 9 \times 9 \times 9$ _____
14. $8 \times 8 \times 8 \times 8 \times 8 \times 8 \times 8$ _____
15. 10×10 _____
16. $10 \times 10 \times 10 \times 10$ _____

Solve.

17. The Forestry Club decided to plant trees in the empty lot behind the football field. They measured the lot and discovered it was 9^2 square yards area. What was the area of the lot?

18. For their next project, the Forestry Club decided to plant a garden in a brick planter in front of the school. The planter can hold 5^3 cubic yards of soil. How many cubic yards of soil can the planter hold?

Use with pages 84–85.

RETEACH: Multiplying Decimals by 10; 100; and 1,000

The school bookstore sold 8.5 cases of erasers. Each case held 100 boxes of erasers. How many boxes did the bookstore sell?

Remember

To multiply by 100, move the decimal two places to the right.

$$8.5 \times 100 = 850.0$$

Multiply money the same way.
To multiply by 10, move one place to the right.
To multiply by 1,000, move three places to the right.

$$10 \times \$1.25 = \$12.50$$
$$100 \times \$1.25 = \$125.00$$
$$1000 \times \$1.25 = \$1,250.00$$

The bookstore sold 850 boxes of erasers.

Multiply.

1. 0.57 × 10
2. 8.94 × 10
3. 0.04 × 10
4. 8.36 × 10
5. 37.503 × 10

6. 233.366 × 10
7. 2.04 × 100
8. 0.0079 × 100
9. 83.88888 × 100
10. 0.0671 × 100

11. 0.0004 × 1000
12. 84.7715 × 1000
13. 5.8 × 1000
14. 2.11 × 1000
15. 0.9 × 1000

Solve.

16. The bookstore had a special sale on dictionaries. They sold 100 dictionaries for $11.75 each. How much money did the bookstore take in for all the dictionaries? _____

RETEACH: Multiplying Decimals by Whole Numbers

A no-frills airline can fly one passenger for $0.0517 per mile.
How much does it cost to fly one passenger 345 miles?

Remember

The product must have the same number of decimal places as the decimal factors.

Multiply	The decimal factor has 4 decimal places.
0.0517	0.(0 5 1 7)—four decimal places
× 345	× 345
2585	17.(8 3 6 5)—four decimal places
2 0680	
15 5100	
17.8365	Round to the nearest cent.

It costs about $17.84 to fly one passenger 345 miles.

Multiply.

1. 4.08 × 7
2. 4.26 × 5
3. 0.6 × 3
4. 4.01 × 71
5. 0.82 × 82

6. 0.0051 × 8
7. 0.0031 × 9
8. 0.0021 × 74
9. 0.04 × 4
10. 0.0015 × 91

11. 88 × 0.0022 = _____
12. 9.48 × 83 = _____
13. 66 × 4.925 = _____

Solve.

14. It costs regular airlines about 10¢ to fly one passenger 1 mile. How much does it cost to fly 473 people 1,000 miles? _____

Use with pages 92–93.

RETEACH: Estimating Decimal Products—Rounding

Sally is going to the grocery store to buy $2\frac{3}{4}$ pounds of bananas. About how much money will she need?

Estimate 2.75 × $0.39

Remember

Round each factor to the largest place.

2.75 ⟶ 3
0.39 ⟶ 0.4
3 × 0.4 = 1.2 or $1.20

Since both factors were rounded up, $1.20 is an overestimate.

The bananas will cost less than $1.20.

Use estimation to place the decimal point.

1. 0.386 × 0.57 = __22002__
2. 5.345 × 7.2 = __38484__
3. 10.51 × 0.816 = __857616__
4. 13.4 × 2.61 = __34974__
5. 1.75 × 0.96 = __1680__
6. 2.975 × 0.134 = __39865__

Estimate. Write > or < for ◯.

7. 2.21 × 3.35 ◯ 6
8. 8 × 3.05 ◯ 24
9. 15.792 × 2.9 ◯ 48
10. 12.17 × 5.29 ◯ 60
11. 6 × 11.8 ◯ 66
12. 9.327 × 0.314 ◯ 3
13. 0.783 × 9.926 ◯ 8
14. 4.175 × 7.4 ◯ 28
15. 0.527 × 0.892 ◯ 1
16. 9 × 7.753 ◯ 72
17. 0.6394 × 15.2 ◯ 9
18. 2.18 × 0.893 ◯ 1

Estimate.

19. 3.95 × 0.27
20. 15.26 × 0.41
21. 28.831 × 2.7
22. 0.871 × 0.3265

RETEACH: Multiplying by a Decimal

As part of her preseason training for basketball, Kathy has to run 1.2 hours each day. If she runs 3.5 miles per hour, how many miles will she run each day?

Remember

First multiply the factors. Then place the decimal point so that the product has as many places as the sum of the decimal places in the factors.

```
    1.2  ← 1 decimal place
  × 3.5  ← 1 decimal place
  -----
    6 0
   36
  -----
   4.20  ← 2 decimal places
```

Kathy will run 4.2 miles each day.

Multiply.

1. 41.3 × 0.2
2. 9.36 × 2.73
3. 4.97 × 5.4
4. 9.69 × 6.1

5. 0.848 × 4.14
6. 0.809 × 2.2
7. 0.906 × 0.4
8. 0.02 × 0.07

9. 0.007 × 0.06
10. 0.311 × 0.5
11. 0.203 × 0.42
12. 0.621 × 0.81

13. $0.51 × 45 = _____
14. $9.84 × 5.2 = _____
15. $931.24 × 0.05 = _____

Solve.

16. On weekends, Kathy runs for 1.5 hours on Saturday and for 0.75 hours on Sunday. If she runs 3.5 miles per hour, how many miles does she run each weekend?

_____

Use with pages 96–97.

RETEACH: Division Facts

The Sundance Outdoor Group sponsored a clean-up rally to remove the empty cans from a local park. At the rally, the group was divided into 3 teams. Each team picked up an equal number of cans. They gathered 24 pounds of cans in all. How many pounds of cans were collected by each team?

Remember

You can divide to find the number of pounds of cans collected by each team. Dividing the pounds of cans into 3 lots is a division problem. You must solve the problem $24 \div 3$.

$$24 \div 3 = 8$$

This division problem can be written in three ways.

$$24 \div 3 = 8 \leftarrow \text{quotient}$$
(dividend ↑, divisor ↑)

$$3\overline{)24} \quad \text{with quotient } 8$$
(dividend, divisor)

$$\frac{24}{3} = 8 \leftarrow \text{quotient}$$
(dividend, divisor)

Each team collected 8 pounds of cans.

Divide.

1. $5\overline{)25}$
2. $8\overline{)8}$
3. $4\overline{)8}$
4. $1\overline{)5}$

5. $9\overline{)0}$
6. $1\overline{)7}$
7. $3\overline{)9}$
8. $3\overline{)3}$

9. $56 \div 8 =$ _____
10. $63 \div 7 =$ _____
11. $30 \div 6 =$ _____
12. $54 \div 9 =$ _____

13. $\frac{14}{7} =$ _____
14. $\frac{36}{6} =$ _____
15. $\frac{32}{4} =$ _____
16. $\frac{81}{9} =$ _____
17. $\frac{72}{8} =$ _____

Solve.

18. To raise money for the Fall Outing, the Sundance Outdoor Group collected acorns. They sold the acorns to a tree farm, where the nuts were used as seeds for growing new trees. All 6 members collected 36 pounds of acorns. If everyone collected the same amount, how many pounds of acorns did each member collect?

RETEACH — Order of Operations

Mark had 3 shelves in his room to display his rock collection. There were 4 rocks on each shelf. For his birthday, Mark's Uncle Alfie gave him 2 more rocks. Write an equation to show how many rocks Mark has in his collection, and solve.

Remember

When parentheses are shown in an equation, always do the operation in parentheses first.

Place parentheses to indicate order of operations and relationship of numbers.

shelves rocks rocks
$(3 \times 4) + 2$
$12 + 2$
14

If the parentheses are moved, the answer is changed.

$(2 + 3) \times 4$
5×4
20

When an expression has two or more operations and there are no parentheses, always multiply or divide from left to right; then add or subtract from left to right. Look at these examples.

$2 + 3 \times 4$ Multiply first. $12 \div 6 - 2$ Divide first.
$2 + 12$ Then add. $2 - 2$ Then subtract.
14 0

Mark had 14 rocks in his collection.

Solve.

1. $35 \div 7 - 2 =$ _____
2. $6 + 42 \div 6 =$ _____
3. $27 - 6 \div 3 =$ _____

4. $32 \div 8 - 4 =$ _____
5. $6 \times 7 - 2 =$ _____
6. $54 + 2 \times 3 =$ _____

7. $15 + 3 \times 4 =$ _____
8. $5 + 6 \times 4 =$ _____
9. $(6 + 42) \div 6 =$ _____

10. $(27 - 6) \div 3 =$ _____
11. $32 \div (8 - 4) =$ _____
12. $35 \div (7 - 2) =$ _____

Solve.

13. As Mark's rock collection grew, he had to add more shelves to display the rocks. He put 3 of his 45 rocks on a table in his room. The rest of the rocks, he divided equally on 6 shelves. Write an equation to show how many rocks he would put on each shelf.

_____

Use with pages 112–113.

RETEACH Divisibility

Carla has 324 buttons in her sewing box. She wants to divide them into 9 separate piles. Can she divide them evenly without having any buttons left?

Remember

There are many rules you can use to determine whether a number is evenly divisible by another number.

Divisor	Rule	Example
2	Any even number is divisible by 2. Even numbers end in 2, 4, 6, 8, or 0. Odd numbers are not divisible by 2.	8; 62; 766; and 1,054 are all divisible by 2.
3	The sum of the digits in the dividend must be divisible by 3.	615 is divisible by 3 because 6 + 1 + 5 = 12, and 12 is divisible by 3.
5	The digit in the ones place must be 0 or 5.	30; 75; 1,245; and 3,560 are all divisible by 5.
6	Any number that is divisible by both 2 and 3 is divisible by 6.	42; 132; 840; and 1,728 are all divisible by 6.
9	The sum of the digits in the dividend must be divisible by 9.	5,328 is divisible by 9 because 5 + 3 + 2 + 8 = 18, and 18 is divisible by 9.
10	The digit in the ones place must be 0.	50; 820; 970; and 1,270 are all divisible by 10.

Yes, Carla can evenly divide the buttons into 9 separate piles.

Each of the numbers below is divisible by one or more of the divisors 2, 3, 5, 6, 9, 10. Write the divisors.

1. 24 _____
2. 16 _____
3. 60 _____
4. 108 _____
5. 369 _____
6. 1,842 _____
7. 58 _____
8. 4,862 _____
9. 14,586 _____
10. 21,879 _____
11. 63,954 _____
12. 3,187,041 _____

Solve.

13. Can 138,567 buttons be evenly divided into 9 equal piles? into 3 equal piles?
_____

RETEACH — Estimating Quotients

Larry's French Club is going to sell crêpes at the school fair. His recipe says each batch of crêpe batter will serve 7 people. If 1,500 people are expected at the fair, about how many batches should Larry expect to make?

Remember

Decide the number of digits in the quotient.

Divide the thousands. Think: $7\overline{)1}$. Not enough thousands.
Divide the hundreds. Think: $7\overline{)15}$.

The quotient will have three digits. $7\overline{)1{,}500}$

Think: $7 \times 2 = 14$.
$7 \times 3 = 21$. Too great. So, use 2. $7\overline{)1{,}500}$ → 2

Write zeros for the other digits. $7\overline{)1{,}500}$ → 200

Larry should expect to make more than 200 batches.

Find the number of digits in each quotient.

1. $6\overline{)2{,}875}$ _____
2. $8\overline{)91{,}295}$ _____
3. $5\overline{)198{,}723}$ _____
4. $3\overline{)7{,}146}$ _____
5. $9\overline{)12{,}430}$ _____
6. $23\overline{)73{,}619}$ _____
7. $15\overline{)32{,}025}$ _____
8. $7\overline{)815{,}436}$ _____
9. $5\overline{)43{,}281}$ _____

Estimate.

10. $6\overline{)5{,}520}$
11. $7\overline{)28{,}955}$
12. $5\overline{)73{,}015}$

13. $3\overline{)169{,}235}$
14. $9\overline{)64{,}413}$
15. $15\overline{)31{,}600}$

16. $23\overline{)475{,}350}$
17. $36\overline{)7{,}318}$
18. $29\overline{)21{,}340}$

19. $8\overline{)57{,}914}$
20. $13\overline{)41{,}639}$
21. $6\overline{)49{,}246}$

Use with pages 116–117.

RETEACH — Dividing by a 1-Digit Divisor

Jill has 324 guppies in a fish tank. She wants to put an equal number of them in 6 smaller tanks. How many guppies will there be in each tank?

Remember

First we divide the hundreds, then the tens, and then the ones.

Divide the hundreds.
Think: 6)3.
Not enough hundreds.

Divide the tens.
Think: 6)30.

```
   5
6)324
  30↓
   24
```
Multiply. Subtract. Compare.

Divide the ones.
Think: 6)24.

```
   54
6)324
  30↓
   24
   24
    0
```
Multiply. Subtract. Compare.

We have 5 tens and 4 ones, or 54.

There will be 54 guppies in each tank.

Divide.

1. 3)48
2. 5)65
3. 4)56
4. 6)72
5. 7)84
6. 5)43

7. 901 ÷ 2 = _____
8. 516 ÷ 6 = _____
9. 400 ÷ 7 = _____

10. $\frac{249}{6}$ = _____
11. $\frac{638}{7}$ = _____
12. $\frac{73}{3}$ = _____
13. $\frac{314}{5}$ = _____

Solve.

14. When Jill buys goldfish for her aquarium, 1 goldfish of every 7 is black instead of gold. How many black goldfish would Jill receive in a shipment of 686 goldfish?

RETEACH: Dividing Larger Numbers

Margaret likes to play a word game called Garble. She scored 1,020 in her last 5 games. On the average, how many points did she score per game?

Remember

You can use zeros to keep place value when you divide.

Divide the thousands.
Divide the hundreds.
Think: 5)10.

Think: 5)1.
Divide the tens.
Think: 5)2.

Not enough thousands.
Divide the ones.
Think: 5)20

```
      2                    2 0                  2 0 4
  5)1,0 2 0            5)1,0 2 0            5)1,0 2 0
    1 0                  1 0                  1 0
      0                    2                    2
                           0                    0
                           2                    2 0
                                                2 0
                                                  0
```

Margaret averaged 204 points per game.

Divide.

1. 5)1,155   2. 6)2,472   3. 8)4,248   4. 6)4,512

5. 4)2,540   6. 9)3,882   7. 6)2,627   8. 4)3,554

Solve.

9. During one string of games, the Jumpers scored 1,035 points, 5 times as many as their best player scored alone. How many points did their best player score?

Use with pages 120–121.

RETEACH Short Division

Vince and his grandfather saw a flock of geese nesting in a marsh. They counted a total of 135 geese in the flock. They noted that one of every 5 geese was a snow goose. How many snow geese were there in the flock?

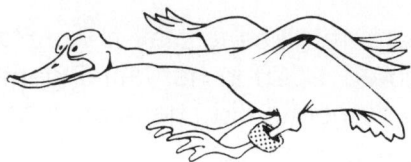

Remember

Sometimes, when you divide by a 1-digit number, it is possible to do the multiplication and subtraction in your head. This is called *short division*.

Divide the hundreds.	Divide the tens.	Divide the ones.
How many 5's in 1?	How many 5's in 13?	How many 5's in 35?
5)135	② 5)1 3 ③ 5	2 ⑦ 5)1 3 ③ 5
Not enough hundreds.	Write the remainder next to the ones.	

There are 27 snow geese.

Divide.

1. 6)624   2. 5)565   3. 7)196   4. 3)651

5. 6)258   6. 2)109   7. 3)404   8. 9)847

9. 6)$42.78   10. 8)$52.16   11. 3)$88.71

Solve.

12. About 900 giant Canada geese spend the winter in Fergus Falls, Minnesota. One of every 6 of these geese has been banded. How many of these geese have been banded?

Use with pages 124–125.

RETEACH: Dividing by Multiples of 10 and 100

Mr. Ruzicka harvested 95,000 pounds of sunflower seeds from 100 acres of land. How many pounds of seeds per acre did he harvest?

Remember

To divide a number by 10, move the decimal point one place to the left. To divide by 100, move the decimal point two places to the left. To divide 95,000 by 100, move the decimal point in 95,000 two places to the left.

95,0.00.

Mr. Ruzicka harvested 950 pounds per acre.

Divide.

1. 10)54
2. 100)54
3. 10)83
4. 100)83

5. 30)27
6. 300)27
7. 20)520
8. 200)520

9. 27,000 ÷ 30 = _____
10. 27,000 ÷ 100 = _____
11. 24,000 ÷ 60 = _____

Solve.

12. The Beaudry farm raises potatoes. Last year, they harvested 11,500 bushels of potatoes from 100 acres of land. How many bushels did they reap per acre?

_____

Use with pages 128–129.

RETEACH: Dividing by a 2-Digit Number

At an Independence Day celebration, 207 people ate 23 bushels of corn. Each bushel of corn fed an equal number of people. How many people did each bushel feed?

Remember

To divide 207 by 23, first divide the hundreds, then the tens, and finally the ones.

Divide the hundreds.	Divide the tens.	Divide the ones.
Think: 23)2.	Think: 23)20.	Think: 23)207.
		9 23)207 207 0
Not enough hundreds.	Not enough tens.	9 ones

Each bushel fed 9 people.

Divide.

1. 24)72
2. 36)144
3. 45)405
4. 17)54

5. 19)97
6. 14)72
7. 32)192
8. 46)368

9. 57)460
10. 922 ÷ 61 = _____
11. 94 ÷ 22 = _____

12. $\frac{795}{35}$ = _____
13. $\frac{93}{31}$ = _____
14. $\frac{513}{43}$ = _____
15. $\frac{708}{59}$ = _____

Solve.

16. At an end-of-school party, 27 sixth-grade students drank 108 glasses of fruit juice. How many glasses of fruit juice did each person drink? _____

Use with pages 130–131.

RETEACH: Correcting Estimates

Ms. Wallace makes bead necklaces. Each necklace has 59 beads. If Ms. Wallace has 360 beads, how many necklaces can she make? How many beads will be left?

Remember

If the estimate of a digit in the quotient is too great or too small, it needs to be corrected.

Divide 59)360.

Divide the hundreds. Think: 59)3. Not enough hundreds.

Divide the tens. Think: 59)36. Not enough tens.

Divide the ones.

Think: ⑤9) ㊱ 0 or 5)36.

Estimate ⑦. 59)360
Multiply. Too great. 413

You need to correct the estimate.

Estimate ⑥. 59)360 ⑥R⑥
Multiply. Subtract. 354
Write the remainder. ⑥

Ms. Wallace can make 6 necklaces with 6 beads left.

Find the quotient.

1. 70)247   2. 45)317   3. 99)289   4. 26)433   5. 32)912

6. 38)722   7. 74)114   8. 27)891   9. 25)135   10. 55)476

Use with pages 132–133.

RETEACH: Dividing Thousands

Each month, the school bus that picks up José takes him a total of 1,610 miles. If the bus takes 46 trips per month, how many miles does it travel on each trip?

Remember

When you bring down a remainder, be careful to subtract correctly. Regroup when necessary.

Divide the thousands.	Divide the hundreds.	Divide the tens. Subtract.	Divide the ones.	
46)1,610	46)1,610	3 46)1,610 1 38 23	35 46)1,610 1 38↓ 230 230 0	Think: Multiply. Subtract. Compare.
Not enough thousands.	Not enough hundreds.			

We have 0 thousands, 0 hundreds, 3 tens, and 5 ones, or 35.

The bus travels 35 miles each trip.

Divide.

1. 67)1273
2. 35)1785
3. 65)4810
4. 26)2392

5. 67)3490
6. 26)2030
7. 16)4544
8. 35)6545

9. 5289 ÷ 43 = _____
10. 8721 ÷ 51 = _____
11. $94.25 ÷ 29 = _____

Solve.

12. The school bus that picks up Maria uses $153.30 worth of gasoline each month. If the bus carries 35 passengers, how many dollars' worth of gasoline are used for each passenger in one month? _____

RETEACH: Dividing Large Numbers

The lunchroom at Westwind School used 178,212 cartons of milk last year. How many cartons of milk were used each month during the 12-month year?

Remember

To divide 178,212 by 12, first divide the hundred thousands, then the ten thousands, the thousands, the hundreds, the tens, and the ones.

The division should look like this. →

```
          14,851
     12)178,212
         12↓
         58
         48↓
         10 2
          9 6↓
            61
            60↓
             12
             12
              0
```

Divide the ten thousands.
Divide the thousands.
Divide the hundreds.
Divide the tens.
Divide the ones.

We have 1 ten thousand, 4 thousands, 8 hundreds, 5 tens, and 1 one, or 14,851.

There were 14,851 cartons of milk used each month.

Divide.

1. 31)67,084
2. 13)38,792
3. 46)75,670
4. 52)64,168

5. 49,153 ÷ 19 = _____
6. 70,020 ÷ 67 = _____
7. 53,470 ÷ 38 = _____

Solve.

8. The heating plant in the Westwind School used 615,400 pounds of coal during 98 heating days last year. How many pounds of coal were used for each heating day?

Use with pages 136–137.

RETEACH: Dividing with Zero

Last month, the sixth-grade class collected 5,382 aluminum cans for a school project. If each of the 26 members of the class collected the same number of cans, how many cans did each member collect?

Remember

To divide 5,382 by 26, first divide thousands, then hundreds, tens, and ones. Be sure to keep the numbers in the quotient in the proper place value.

Divide the thousands.	Divide the hundreds.	Divide the tens.	Divide the ones.
26)5,382 Not enough thousands.	2 26)5,382 52 1	20 26)5,382 52↓ 18 0 Not enough tens. Put a zero in the tens place.	207 26)5,382 52↓ 18 0↓ 182 182 0

There are 2 hundreds, 0 tens, and 7 ones, or 207.

Each student collected 207 cans.

Divide.

1. 27)8,154

2. 43)8,729

3. 12)9,684

4. 35)3,745

5. 34)$70.38

6. 74)15,244

7. 30)15,180

8. 97)9,991

9. 7,960 ÷ 73 = _____

10. $131.67 ÷ 63 = _____

11. 17,378 ÷ 57 = _____

Use with pages 138–139.

RETEACH: Dividing by a 3-Digit Divisor

Last year, the county soil-conservation team planted 1,327,900 trees in 245 planting days. How many trees did they plant each day?

Remember

When there is a zero in the ones place of a quotient, always be sure to keep it in the answer.

```
          5,420
245)1,327,900
    1 225
      102 9
       98 0
        4 90
        4 90        No remainder.
           0        Carry the zero to the ones place.
```

We have 5 thousands, 4 hundreds, 2 tens, and 0 ones, or 5,420.

They planted 5,420 trees each day.

Divide.

1. 430)158,670

2. 367)88,447

3. 275)215,600

4. 156)81,744

5. 346)493,396

6. 451)542,102

7. 64,510 ÷ 300 = _____

8. 128,445 ÷ 169 = _____

9. $4,139.22 ÷ 298 = _____

Use with pages 140–141.

RETEACH — Equations

Emil has 36 pairs of sunglasses to display for sale. He has an equal number of 4 different styles. If Emil arranges the glasses by style, how many pairs of sunglasses will there be in each group?

Remember

To find how many in each group, divide the total number by the number of groups.

$$36 \div 4 = n$$
$$36 \div 4 = 9$$

There will be 9 pairs of sunglasses in each group.

Find the missing number.

1. _____ × 20 = 180
2. _____ × 60 = 420
3. _____ × 30 = 180

4. _____ × 80 = 160
5. 7 × _____ = 140
6. 5 × _____ = 150

7. 8 × _____ = 200
8. 3 × _____ = 240
9. _____ ÷ 5 = 15

10. _____ ÷ 2 = 80
11. _____ ÷ 3 = 35
12. _____ ÷ 9 = 27

13. 640 ÷ _____ = 320
14. 15 ÷ _____ = 3
15. 810 ÷ _____ = 90

Solve.

16. Emil has 9 coins of the same denomination that total $2.25. What denominations are they? _____

Use with pages 142–143.

RETEACH: Dividing Decimals by Whole Numbers

Last year, Mr. Feeney gathered 928.2 pounds of walnuts from his 7 walnut trees. How many pounds of walnuts per tree did he gather?

Remember

First estimate the quotient to place the decimal point.

Divide the hundreds.	Divide the tens.	Divide the ones.	Divide the tenths.
1 . 7)928.2 7 — 2	13 . 7)928.2 7↓ — 22 21 — 1	132. 7)928.2 7↓↓ — 22 21↓ — 18 14 — 4	132.6 7)928.2 7↓↓ — 22 21↓ — 18 14↓ — 4.2 4.2 — 0

We have 1 hundred, 3 tens, 2 ones, and 6 tenths, or 132.6.

Mr. Feeney gathered 132.6 pounds of walnuts per tree.

Divide.

1. 7)60.2   2. 3)27.6   3. 8)59.2   4. 6)48.6

5. 8)28.48   6. 5)14.35   7. 4)25.24   8. 2)10.96

Use with pages 154–155.

RETEACH: More Dividing by Whole Numbers

Jed ran a total of 6.25 km in 5 days. What was the average number of kilometers he ran each day?

Remember

Divide the whole number. Place the decimal point in the quotient; then divide the decimal.

```
      1.            1.2           1.25
  5)6.25        5)6.25        5)6.25
    5             5 ↓           5 ↓
    ─             ─             ─
    1             1 2           1 2
                  1 0           1 0 ↓
                  ─             ─
                    2             25
                                  25
                                  ──
                                   0
```

Jed ran an average of 1.25 km each day.

Divide.

1. 3)16.35
2. 4)48.24
3. 7)7.175
4. 6)39

5. 7)$42.35
6. 5)2.3
7. 5)6.26
8. 54)0.27

9. 26 ÷ 8 = _____
10. $54.60 ÷ 5 = _____
11. 80.56 ÷ 4 = _____

Solve.

12. To get ready for a race, Jed works out with a skip rope every day. In 100 days, he puts in 1,227 minutes of skipping rope. How many minutes is this per day?

_____

RETEACH: Dividing Decimals by 10; 100; 1,000

During strawberry season, Scott and Kevin pick an average of 115.5 pints of strawberries for every 10 hours of picking time. On the average, how many pints of strawberries do the boys pick per hour?

Remember

When dividing by 10, move the decimal point one place to the left.
When dividing by 100, move the decimal point two places to the left.
When dividing by 1,000, move the decimal point three places to the left.

115.5 ÷ 10 = 1 1 5 5 = 11.55

The boys average 11.55 pints of strawberries per hour.

Divide.

1. 10)35.6
2. 10)4.27
3. 10)0.81
4. 10)0.09

5. 100)429.7
6. 100)63.92
7. 100)8.903
8. 100)$77.00

9. 1000)5280
10. 1000)980
11. 1000)75.4
12. 1000)2.54

Use with pages 158–159.

RETEACH: Dividing by a Decimal

Smitty operates a shoeshine stand during the summer. The fastest he can shine a shoe is 1.06 minutes. At this rate, how many shoes can he shine in 31.8 minutes?

Remember

Move the decimal points in both the divisor and the dividend an equal number of places to make the divisor a whole number.

Multiply both divisor and dividend by 100. Move each decimal point two places to the right.

$$1.06\overline{)31.80}$$

Place the decimal point in the quotient.

$$106\overline{)3180.}$$

Divide.

$$106\overline{)3180.} \quad \begin{array}{r} 30. \\ \underline{318\downarrow} \\ 0 \end{array}$$

Smitty can shine 30 shoes in 31.8 minutes.

Divide.

1. $0.7\overline{)0.07}$
2. $0.5\overline{)0.39}$
3. $0.2\overline{)0.42}$
4. $0.6\overline{)0.42}$

5. $0.3\overline{)266.1}$
6. $0.2\overline{)0.718}$
7. $8.7\overline{)0.8265}$
8. $0.5\overline{)0.004}$

9. $0.024 \div 0.6 =$ _____
10. $8.7 \div 0.3 =$ _____
11. $23.82 \div 0.3 =$ _____

Solve.

12. Jack's shoe-shine stand earned $111.30 in 63.6 hours of operation. How much money did Jack earn for each hour of operation? _____

RETEACH: More Dividing by a Decimal

After spending 6.5 days at sea, a fishing boat returned with 20.8 tons of fish. On the average, how many tons of fish did they catch per day?

Remember

To find how many tons per day, you must divide the total tons by the number of days.

$$\frac{\text{total tons}}{\text{number of days}} = \frac{20.8}{6.5}$$

Multiply the divisor and dividend by 10 to move the decimal point one place to the right.

$$6.5\overline{)20.8}$$

Place the decimal point in the quotient.

$$65\overline{)208.}$$

Divide. Write zero in the remainder when necessary.

$$\begin{array}{r} 3.2 \\ 65\overline{)208.0} \\ \underline{195}\downarrow \\ 130 \leftarrow \text{Add a zero} \\ \underline{130} \\ 0 \end{array}$$

They caught an average of 3.2 tons of fish per day.

Divide.

1. $3.8\overline{)10.26}$
2. $4.3\overline{)27.95}$
3. $1.1\overline{)0.22}$
4. $8.3\overline{)2.905}$

5. $26.144 \div 8.6 =$ _____
6. $53.94 \div 6.2 =$ _____
7. $2.438 \div 0.53 =$ _____

Solve.

8. Mr. Kemph's fishing boat returned to port with 10.2 tons of fish. Mr. Kemph sold the fish to a cannery for $7,650. How much did Mr. Kemph receive per ton of fish?

_____

Use with pages 162–163.

RETEACH Rounding Decimal Quotients

A kite string is wound on a spool so that 1 turn of the spool lets out 1.75 feet of line. How many turns of the spool will be needed to let out 1,275 feet of line?

Round your answer to the nearest whole turn.

Remember

To round to the nearest whole number, look at the tenths place.

```
            728.5
     1.75)1275.0 0 0
          1225
           500
           350
           1500
           1400
            1000
             875
             125
```

Round up if tenths are 5 or more.
Round down if tenths are less than 5.

It takes about 729 turns to let out all the string.

Round to the nearest tenth.

1. 57.47 _____
2. 13.12 _____
3. 91.99 _____
4. 19.80 _____

5. 84.30 _____
6. 78.18 _____
7. 93.15 _____
8. 29.30 _____

Round to the nearest hundredth.

9. 2.435 _____
10. 4.351 _____
11. 2.164 _____
12. 5.678 _____

13. 3.535 _____
14. 1.777 _____
15. 4.123 _____
16. 4.911 _____

Round to the nearest thousandth.

17. 4.3257 _____
18. 5.1346 _____
19. 6.1435 _____
20. 6.5793 _____

21. 5.3535 _____
22. 9.1999 _____
23. 3.2140 _____
24. 7.1418 _____

RETEACH: Metric Units of Length

Many different things, such as Olympic events and photography, are measured in metric units. In 35-millimeter film, the *35* is the length of the negative in millimeters. An Olympic-size track is 400 meters long.

Remember

These are the relationships of metric units.

10 millimeters (mm) = 1 centimeter (cm)
10 centimeters (cm) = 1 decimeter (dm)
10 decimeters (dm) = 1 meter (m)
1,000 meters (m) = 1 kilometer (km)

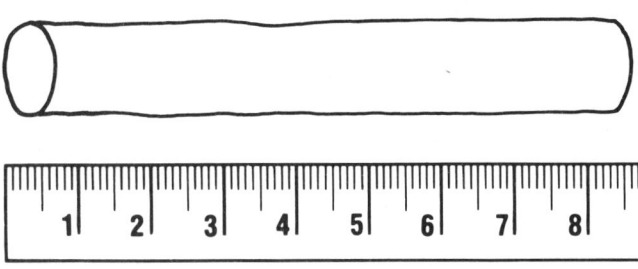

The length of the chalk to the nearest centimeter is 9 cm.
The length of the chalk to the nearest millimeter is 86 mm.
Note that 86 mm = 8.6 cm.

Write the unit you would use to measure

1. the length of this page. _____

2. the height of your school building. _____

3. the length of a soccer field. _____

4. the distance from your home to school. _____

5. the thickness of a dime. _____

6. the thickness of a shirt button. _____

7. the length of your thumbnail. _____

8. the length of a semitrailer. _____

9. the distance in a marathon race. _____

10. the diameter of a tennis ball. _____

Use with pages 170–171.

RETEACH: Equal Metric Measures of Length

If Anita runs a 30,000-meter race, how many kilometers does she run?

The relationships of the metric units are:
10 millimeters (mm) = 1 centimeter (cm)
10 centimeters (cm) = 1 decimeter (dm)
10 decimeters (dm) = 1 meter (m)
1000 meters (m) = 1 kilometer (km)

Remember
To rename smaller units with larger units, divide. To rename larger units with smaller units, multiply. To rename meters (smaller units) with kilometers (larger units), divide by 1,000.

$$30,000 \div 1,000 = 30$$

Anita runs 30 km.

Complete.

1. 0.35 mm = _____ cm
2. 3,000 m = _____ km
3. 9.6 km = _____ m
4. 1.5 m = _____ cm
5. 390 m = _____ cm
6. 0.315 km = _____ m
7. 5,000 m = _____ km
8. 274 cm = _____ m
9. 0.9042 m = _____ km
10. 0.678 km = _____ m
11. 96.74 km = _____ m
12. 36,106 cm = _____ m
13. 53 dm = _____ m
14. 19 m = _____ dm
15. 152 cm = _____ dm

Solve.

16. How many kilometers long is a 4,000-meter race?

_____

Use with pages 172–173.

RETEACH: Metric Units of Capacity and Mass

Marie bought a bag of catfood for her cat. The bag weighs 1.81 kg. How many grams is this?

Remember

Liters measure capacity, and grams measure mass.

1,000 milliliters (mL) = 1 liter (L)
1,000 liters (L) = 1 kiloliter (kL)

1,000 milligrams (mg) = 1 gram (g)
1,000 grams (g) = 1 kilogram (kg)

To rename larger units with smaller units, multiply. To rename smaller units with larger units, divide. To rename kilograms (larger units) with grams (smaller units), multiply by 1,000.

$1.81 \times 1,000 = 1,810$

There are 1,810 grams of catfood in the bag.

Choose the correct unit you would use to measure each.

1. a cat _____
2. a bug _____
3. a flake of snow _____
4. a semitrailer _____
5. a toy airplane _____
6. your shoes _____
7. a drop of water _____
8. a cup of coffee _____
9. a bottle of medicine _____
10. a soft-drink can _____
11. a gasoline can _____
12. a perfume bottle _____

Complete.

13. 400 g = _____ kg
14. 2.5 L = _____ kL
15. 5 kL = _____ L
16. 2,500 kg = _____ g
17. 45,000 L = _____ kL
18. 45 mL = _____ L
19. 35 mg = _____ g
20. 99 g = _____ mg
21. 1 kL = _____ L

Use with pages 176–177.

RETEACH   Least Common Multiples

Nancy runs 6 miles each day. Robin runs 4 miles each day. What is the fewest days each must run before they have both run the same number of miles?

> **Remember**
>
> To find the fewest days each girl must run, find the least common multiple of 6 and 4.
>
> Multiples of 6: 0, 6, 12, 18, 24, 30, 36, 42. . . .
> Multiples of 4: 0, 4, 8, 12, 16, 20, 24, 28. . . .
>
> The common multiples of 6 and 4 are 0, 12, 24. . . . Zero is not considered to be a least common multiple. Think: $6 \times 2 = 12$, and $4 \times 3 = 12$. The least common multiple of 6 and 4 is 12.

Nancy must run for 2 days and Robin must run for 3 days before both have run 12 miles.

List the first six multiples of each number.

1. 3
2. 4
3. 7
4. 9
5. 6
6. 5
7. 10
8. 12
9. 15

Find the least common multiple of each pair of numbers.

10. 3 and 4 _____
11. 6 and 8 _____
12. 9 and 12 _____
13. 18 and 24 _____
14. 5 and 12 _____
15. 7 and 5 _____

Solve.

16. On a basketball floor, Jeff can make 9 free throws in 5 minutes. Nathan can make 12 free throws in 5 minutes. At this rate, what is the least amount of time it will take each of them to shoot the same number of free throws?

_____

54   Use with pages 190–191.

RETEACH   Greatest Common Factor

Hans caught 24 fish, and Bob caught 18 fish. If they both caught the same number of fish per hour, how many fish did they catch per hour? What is the least number of hours that each could have fished?

Remember

To find the number of fish per hour each caught, find the greatest common factor of 24 and 18.

Factors of 24: 1, 2, 3, 4, 6, 8, 12, and 24.
Factors of 18: 1, 2, 3, 6, 9, and 18.

The common factors of 18 and 24 are 1, 2, 3, and 6. The greatest common factors of 18 and 24 is 6.

Each caught 6 fish per hour. Hans fished for 4 hours and Bob fished for 3 hours.

List the factors.

1. 6 _____   2. 9 _____   3. 5 _____   4. 2 _____   5. 8 _____

List the common factors of each pair of numbers.

6. 6 and 8 _____   7. 5 and 15 _____   8. 8 and 12 _____

Write the greatest common factor of each pair of numbers.

9. 6 and 8 _____   10. 5 and 15 _____   11. 8 and 12 _____

Solve.

12. Joanie caught 15 fish, and Jennifer caught 25 fish. If each caught the same number of fish per hour, what is the least number of hours each could have fished? How many fish per hour did they catch?

_____

_____

Use with pages 192–193.

RETEACH — Primes, Composites, Prime Factors

Their father built Skipper and Janie a new clubhouse. The area of the clubhouse is 221 square feet. What are the length and the width of the club house?

> **Remember**
>
> Find two numbers whose product is 221. Since each composite number has only one set of prime factors, begin by dividing 221 by the prime factors 2, 3, 5, 7, 11, 13, 17, and so on, until an exact divisor or factor is found and there is no remainder.
>
> 221 ÷ 2 = 110 + R1
> 221 ÷ 3 = 73 + R2
> 221 ÷ 5 = 44 + R1
> 221 ÷ 7 = 31 + R4
> 221 ÷ 11 = 20 + R1
> 221 ÷ 13 = 17   No remainder.
> 221 ÷ 13 = 17; so, 17 × 13 = 221.

The club house is 17 feet by 13 feet.

Write the factors of each number. Write whether the number is *prime* or *composite*.

1. 5
2. 8
3. 6
4. 12

5. 20
6. 15
7. 21
8. 16

9. 23
10. 27
11. 51
12. 33

Solve.

13. Skipper built a new rabbit pen. The area of the floor is 572 square feet. What length and width could the pen have and yet most nearly resemble a square?

Use with pages 194–195.

RETEACH: Fractions and Equivalent Fractions

Oscar has a garden that is divided into five sections. Two sections are used for strawberries. Three sections are used for vegetables. If the garden were divided into ten sections, what fraction of the garden would be used for vegetables?

Strawberries	Strawberries	Vegetables	Vegetables	Vegetables

Remember

Two fifths of the garden is strawberries; three fifths is for vegetables.

$$\frac{3}{5} = \frac{3 \times 2}{5 \times 2} = \frac{6}{10}$$

Six tenths of the garden is used for vegetables.

Write the fraction for the shaded part of each shape.

1.
2.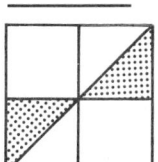
3.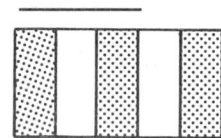
4.

5.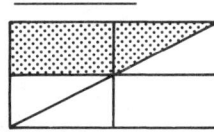
6.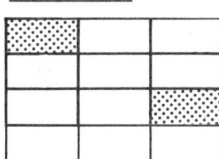
7.
8.

Solve.

9. Sarah has 8 pumpkins left at her stand. She has promised 6 of them to neighbors. What fraction of the pumpkins remain to be sold?

10. Oscar grew 20 pints of strawberries. He promised to give 5 pints to his best friend, Sallianne. What fraction of the strawberries will he have left to sell?

Use with pages 198–199.

RETEACH: Simplifying Fractions

On the archery range, two members of the school team hit 12 bulls-eyes in 14 targets. At this rate, how many bulls-eyes would they hit in 7 targets?

Remember
To simplify a fraction, divide both numerator and denominator by the same factor.
$$\frac{12}{14} = \frac{12 \div 2}{14 \div 2} = \frac{6}{7}$$

Six bulls-eyes would be hit.

Write the fraction in simplest form.

1. $\frac{3}{6} =$ _____
2. $\frac{4}{8} =$ _____
3. $\frac{4}{6} =$ _____
4. $\frac{5}{10} =$ _____
5. $\frac{6}{8} =$ _____
6. $\frac{15}{20} =$ _____
7. $\frac{3}{9} =$ _____
8. $\frac{12}{18} =$ _____
9. $\frac{24}{36} =$ _____
10. $\frac{40}{60} =$ _____

Complete.

11. $\frac{6}{8} = \frac{\_\_}{4}$
12. $\frac{8}{12} = \frac{4}{\_\_}$
13. $\frac{\_\_}{24} = \frac{10}{12}$
14. $\frac{\_\_}{40} = \frac{15}{20}$

Solve.

15. In one archery match, 16 of the 18 participants hit bulls-eyes. At this rate, how many bulls-eyes would be hit if there were 9 participants?

_____

RETEACH Writing Decimals for Fractions

During one part of the baseball season, Pete's batting average was $\frac{38}{125}$, or 38 hits for 125 times at bat. What was his batting average in decimals?

Remember

Change $\frac{38}{125}$ to a decimal by dividing.

$$38 \div 125 = 0.304 \quad \text{or} \quad \begin{array}{r} 0.304 \\ 125\overline{)38.000} \\ \underline{37\ 5} \\ 500 \\ \underline{500} \\ 0 \end{array}$$

Pete's batting average was 0.304.

Write as a decimal. Round the answer to the nearest hundredth.

1. $\frac{3}{10}$ _____
2. $\frac{6}{10}$ _____
3. $\frac{5}{10}$ _____
4. $\frac{1}{2}$ _____
5. $\frac{3}{4}$ _____
6. $\frac{5}{7}$ _____
7. $\frac{5}{9}$ _____
8. $\frac{1}{6}$ _____
9. $\frac{7}{11}$ _____
10. $\frac{1}{7}$ _____

Solve.

11. Ty Cobb had 4,191 hits and was at bat 11,420 times before he retired. What was his batting average? _____

12. When Whitney joined the Little League baseball team, she could hit the ball 13 times for every 50 times she went to bat. She practiced very hard, and after a few months, her batting average improved to 0.320 for every 50 times she went to bat. What was Whitney's batting average when she first started? After her average improved, how many times could she hit the ball for every 50 times at bat? _____

Use with pages 202–203.

RETEACH: Mixed Numbers and Fractions

Kim had a slumber party and ordered 2 large pizzas. Each pizza was cut into pieces as shown here. The shaded part was eaten. Write a fraction to show how much of the 2 pizzas was eaten.

Remember

Since each pizza is cut into 8 pieces, each piece is $\frac{1}{8}$. So 14 pieces were eaten, or $\frac{14}{8}$.

Write the fraction in simplest form. $\frac{14}{8} = \frac{7}{4} = 1\frac{3}{4}$

The girls ate $1\frac{3}{4}$ pizzas.

Write a fraction or mixed number to describe the shaded region.

1. _____
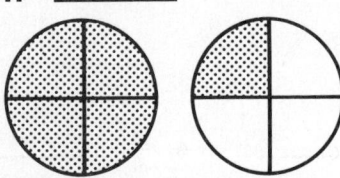

2. _____
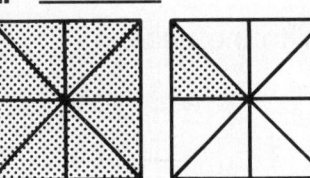

3. _____
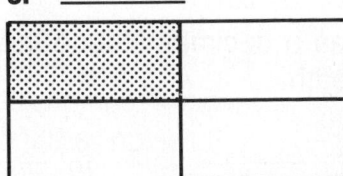

4. _____
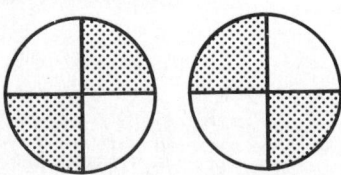

5. _____

6. _____
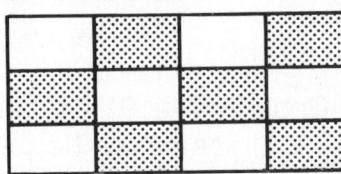

Write as a whole number or a mixed number.

7. $\frac{4}{3} =$ _____
8. $\frac{18}{4} =$ _____
9. $\frac{14}{6} =$ _____
10. $\frac{7}{4} =$ _____
11. $\frac{13}{3} =$ _____

Solve.

12. After Kim and her friends had feasted on pizza, they decided it was time for frozen yogurt. Kim ordered 2 cartons of her favorite flavor, red razzleberry rocky road. Each carton contains 9 servings, and Kim and her friends ate all but 2 servings. Write a fraction to show how much they ate.

_____

RETEACH: Comparing and Ordering Fractions

After a tea, Dawn and Beth had to wash the cups and saucers. Dawn washed $\frac{2}{3}$ of the cups, while Beth washed $\frac{5}{8}$ of the saucers. Which girl washed more pieces?

Remember

In order to compare fractions, the fractions must have like denominators. Find the least common multiple of 3 and 8.

$$\frac{2}{3} = \frac{2 \times 8}{3 \times 8} = \frac{16}{24}$$

$$\frac{5}{8} = \frac{5 \times 3}{8 \times 3} = \frac{15}{24}$$

$$\frac{16}{24} > \frac{15}{24}; \text{ so, } \frac{2}{3} > \frac{5}{8}$$

Dawn washed more pieces.

Compare. Write $>$, $<$, or $=$.

1. $\frac{4}{9}$ _____ $\frac{5}{9}$
2. $\frac{3}{5}$ _____ $\frac{2}{5}$
3. $4\frac{5}{8}$ _____ $4\frac{3}{8}$
4. $7\frac{2}{5}$ _____ $7\frac{3}{5}$

5. $\frac{3}{4}$ _____ $\frac{1}{2}$
6. $\frac{1}{3}$ _____ $\frac{3}{8}$
7. $\frac{3}{8}$ _____ $\frac{4}{7}$
8. $\frac{1}{6}$ _____ $\frac{1}{7}$

9. $\frac{4}{3}$ _____ $1\frac{2}{3}$
10. $3\frac{1}{2}$ _____ $\frac{7}{2}$
11. $4\frac{1}{3}$ _____ $\frac{15}{3}$
12. $\frac{7}{4}$ _____ $1\frac{3}{8}$

Solve.

13. At a reception for new teachers, $1\frac{3}{4}$ hams and $1\frac{7}{8}$ roasts were eaten. Did the guests eat more hams or more roasts?

_____

14. At a baby shower, Julie's aunt used $1\frac{5}{6}$ pounds cheese to make appetizers. She served $1\frac{2}{3}$ pounds cashews. Did Julie's aunt use more cheese or more cashews?

_____

Use with pages 206–207.

RETEACH: Adding Fractions

Scott traded $\frac{1}{5}$ of his baseball-card collection for a baseball. He traded $\frac{1}{3}$ of his baseball-card collection for a football. What fraction of his card collection has Scott traded?

Remember

To add fractions that have unlike denominators, you need to write the fractions as equivalent fractions that have like denominators.

$$\frac{1}{5} + \frac{1}{3} = \frac{1 \times 3}{5 \times 3} + \frac{1 \times 5}{3 \times 5}$$
$$= \frac{3}{15} + \frac{5}{15}$$
$$= \frac{8}{15}$$

Scott has traded $\frac{8}{15}$ of his collection.

Add. Write the answer in simplest form.

1. $\frac{2}{5} + \frac{3}{5} = $ _____
2. $\frac{2}{7} + \frac{4}{7} = $ _____
3. $\frac{1}{6} + \frac{4}{10} = $ _____

4. $\frac{5}{12} + \frac{7}{12} = $ _____
5. $\frac{1}{10} + \frac{4}{10} = $ _____
6. $\frac{1}{3} + \frac{3}{4} = $ _____

7. $\frac{3}{4} + \frac{1}{8} = $ _____
8. $\frac{4}{9} + \frac{5}{9} = $ _____
9. $\frac{1}{4} + \frac{1}{5} = $ _____

10. $\frac{5}{8} + \frac{2}{8}$
11. $\frac{2}{5} + \frac{3}{10}$
12. $\frac{3}{4} + \frac{5}{16}$
13. $\frac{2}{3} + \frac{1}{5}$
14. $\frac{3}{20} + \frac{2}{30}$

Solve.

15. Katie got $\frac{1}{3}$ of her rock collection from her brother. She got $\frac{2}{5}$ of her collection from her uncle. She found the other rocks. What part of the collection was given to Katie?

RETEACH: Subtracting Fractions

During the summer, Mary spends $\frac{1}{5}$ of her day on her paper route. During the school year, she spends only $\frac{1}{8}$ of her day on the paper route. Write a fraction to show how much more of her day she spends on her paper route during the summer.

Remember

To subtract fractions that have unlike denominators, first find a common denominator.

$$\frac{1}{5} - \frac{1}{8} = \frac{1 \times 8}{5 \times 8} - \frac{1 \times 5}{8 \times 5}$$

$$= \frac{8}{40} - \frac{5}{40}$$

$$= \frac{3}{40}$$

Mary spends $\frac{3}{40}$ more time on her paper route during the summer.

Subtract. Write the difference in simplest form.

1. $\frac{3}{5} - \frac{1}{5} =$ _____
2. $\frac{5}{8} - \frac{3}{8} =$ _____
3. $\frac{7}{10} - \frac{5}{10} =$ _____

4. $\frac{4}{6} - \frac{1}{6} =$ _____
5. $\frac{7}{10} - \frac{1}{4} =$ _____
6. $\frac{5}{8} - \frac{1}{4} =$ _____

7. $\frac{5}{6} - \frac{2}{3} =$ _____
8. $\frac{1}{2} - \frac{3}{8} =$ _____
9. $\frac{5}{12} - \frac{1}{4} =$ _____

Solve.

10. Mary spends $\frac{1}{3}$ of her time sleeping and $\frac{2}{5}$ of her time in school. What part of her day is spent doing other things?

_____

Use with pages 212–213.

RETEACH: Adding Mixed Numbers

Jill's father painted $5\frac{1}{2}$ houses in June and $4\frac{1}{3}$ houses in July. How many houses did he paint during the two-month period?

Remember
Find a common denominator. Add the fractions. Then add the whole numbers.

Find the equivalent fractions that have common denominators.

$5\frac{1}{2} \rightarrow 5\frac{3}{6}$
$4\frac{1}{3} \rightarrow 4\frac{2}{6}$

Add the fractions.

$5\frac{3}{6}$
$+\,4\frac{2}{6}$
$\overline{\frac{5}{6}}$

Add the whole numbers.

$5\frac{3}{6}$
$+\,4\frac{2}{6}$
$\overline{9\frac{5}{6}}$

Jill's father painted $9\frac{5}{6}$ houses.

Add. Write the sum in simplest form.

1. $2\frac{1}{4} + 5\frac{2}{4} =$ _____
2. $3\frac{3}{5} + 4\frac{1}{5} =$ _____
3. $5\frac{3}{10} + 6\frac{3}{10} =$ _____

4. $12\frac{1}{6}$
   $+\ 5\frac{3}{6}$

5. $7\frac{1}{3}$
   $+\ 2\frac{1}{3}$

6. $3\frac{1}{6}$
   $+\ 4\frac{2}{3}$

7. $4\frac{3}{8}$
   $+\ 1\frac{1}{8}$

8. $5\frac{2}{3}$
   $+\ 6\frac{1}{4}$

9. $2\frac{1}{5}$
   $+\ 8\frac{3}{4}$

10. $2\frac{1}{2}$
    $+\ 7\frac{1}{4}$

11. $8\frac{1}{5}$
    $+\ 6\frac{2}{10}$

Solve.

12. It took Jill's dad $19\frac{3}{4}$ hours to paint one house and $14\frac{1}{2}$ hours to paint another house. How many hours did it take to paint the two houses?

_____

RETEACH: Subtracting Mixed Numbers

Rita spends $9\frac{3}{4}$ hours per week helping around the house. She spends $6\frac{1}{2}$ hours of this time in the kitchen and the rest of the time working in the yard. How much time does she spend working in the yard?

Remember

Find equivalent mixed numbers that have like denominators. Then subtract.

Find the fractions that have like denominators.	Subtract the fractions.	Subtract the whole numbers.
$9\frac{3}{4} \rightarrow 9\frac{3}{4}$ $6\frac{1}{2} \rightarrow 6\frac{2}{4}$	$9\frac{3}{4}$ $-6\frac{2}{4}$ $\frac{1}{4}$	$9\frac{3}{4}$ $-6\frac{2}{4}$ $3\frac{1}{4}$

Rita spends $3\frac{1}{4}$ hours working in the yard.

Subtract. Write the answer in simplest form.

1. $6\frac{2}{3} - 4\frac{1}{3}$
2. $7\frac{5}{6} - 3\frac{1}{6}$
3. $6\frac{7}{8} - 2\frac{1}{8}$
4. $6\frac{9}{10} - 3\frac{2}{10}$
5. $9\frac{5}{12} - 5\frac{1}{12}$

6. $8\frac{1}{2} - 6\frac{1}{4}$
7. $6\frac{1}{3} - 2\frac{1}{6}$
8. $7\frac{3}{10} - 5\frac{1}{5}$
9. $12\frac{2}{3} - 6\frac{7}{12}$
10. $6\frac{1}{5} - 4\frac{2}{15}$

Solve.

11. Carlos baby-sits for $24\frac{3}{4}$ hours in one month. He is paid for $12\frac{1}{2}$ hours. How many hours does he still have to be paid for?

12. To earn her allowance, Juanita helps her father in his garden. She works 7 hours each week. If she has already worked $3\frac{3}{4}$ hours this week, how much longer must she work to earn her allowance?

Use with pages 216–217.

RETEACH: Adding and Subtracting with Renaming

Gerald and Gary have a lawn-cutting business. They estimated that a certain job would take them $4\frac{1}{4}$ hours to complete. They completed the job in $2\frac{1}{2}$ hours. How many hours did they save?

Remember

Sometimes you need to rename in order to subtract.

Find fractions that have like denominators.	Compare fractions. Rename.	Subtract.
$4\frac{1}{4} \rightarrow 4\frac{1}{4}$ $-2\frac{1}{2} \rightarrow -2\frac{2}{4}$	$4\frac{1}{4} = 3 + 1\frac{1}{4}$ $= 3\frac{5}{4}$	$4\frac{1}{4} \rightarrow 3\frac{5}{4}$ $-2\frac{2}{4} \rightarrow 2\frac{2}{4}$ $\overline{1\frac{3}{4}}$

They saved $1\frac{3}{4}$ hours.

Add. Write the sum in simplest form.

1. $2\frac{1}{4} + 3\frac{3}{4} = $ _____

2. $4\frac{3}{8} + 5\frac{7}{8} = $ _____

3. $2\frac{5}{6} + 7\frac{4}{6} = $ _____

Subtract. Write the difference in simplest form.

4. $7\frac{5}{11} - 5\frac{8}{11} = $ _____

5. $8\frac{1}{5} - 4\frac{2}{5} = $ _____

6. $8\frac{2}{6} - 5\frac{4}{6} = $ _____

Add or subtract. Write the answer in simplest form.

7. $7\frac{1}{4}$
   $-3\frac{1}{2}$

8. $8\frac{6}{10}$
   $+7\frac{2}{5}$

9. $7\frac{3}{8}$
   $-3\frac{1}{2}$

10. $6\frac{3}{8}$
    $+9\frac{3}{4}$

Solve.

11. Gerald and Gerry estimated that it would take them $5\frac{3}{4}$ hours to trim a hedge. It actually took them $7\frac{1}{2}$ hours. How much more time did it take than they had estimated?

Use with pages 218–219.

RETEACH: Multiplying Fractions by Fractions

Of the students in the sixth grade at Midway School, $\frac{1}{2}$ have bicycles. Of these students, $\frac{3}{4}$ have blue bicycles. What part of the class has blue bicycles?

Remember

To find $\frac{3}{4}$ of $\frac{1}{2}$, draw a picture.

$\frac{1}{2}$ of the students

$\frac{3}{4}$ of $\frac{1}{2}$ of the students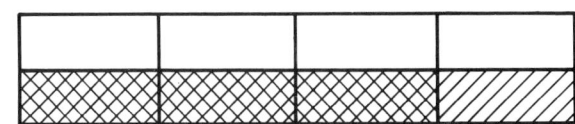

You can multiply the fractions without drawing pictures.

$\frac{3}{4}$ of $\frac{1}{2} = \frac{3}{4} \times \frac{1}{2} = \frac{3 \times 1}{4 \times 2} = \frac{3}{8}$ (multiply numerators) (multiply denominators)

At Midway School, $\frac{3}{8}$ of the class has blue bicycles.

Multiply.

1. $\frac{1}{6} \times \frac{1}{5} =$ _____
2. $\frac{1}{4} \times \frac{1}{3} =$ _____
3. $\frac{1}{2} \times \frac{1}{2} =$ _____
4. $\frac{3}{5} \times \frac{1}{4} =$ _____
5. $\frac{1}{6} \times \frac{3}{4} =$ _____
6. $\frac{1}{3} \times \frac{3}{4} =$ _____
7. $\frac{2}{3} \times \frac{5}{8} =$ _____
8. $\frac{2}{3} \times \frac{3}{4} =$ _____
9. $\frac{2}{5} \times \frac{5}{8} =$ _____
10. $\frac{4}{14} \times \frac{7}{8} =$ _____
11. $\frac{2}{3} \times \frac{1}{5} =$ _____
12. $\frac{1}{2} \times \frac{6}{7} =$ _____

Solve.

13. One third of the children in the sixth grade at Midway School have pets. One half of these students have dogs. What part of the sixth-grade children at Midway have dogs?

14. Of all the children with dogs at Midway School, half have collies. Of these collies, one third are still puppies. What fraction of the children with dogs have collie puppies?

Use with pages 230–231.

RETEACH: Multiplying Fractions and Whole Numbers

Students in the sixth grade at West School spent 14 hours one week at part-time jobs. Of these hours, $\frac{1}{5}$ were spent baby-sitting. How many of the 14 working hours were spent baby-sitting?

Remember

To multiply a fraction and a whole number, think of the whole number as a fraction also.

$\frac{1}{5} \times \frac{14}{1}$ Write 14 as $\frac{14}{1}$.

$\frac{1 \times 14}{5 \times 1} = \frac{14}{5}$ Multiply the fractions.

$\frac{14}{5} = 2\frac{4}{5}$ Simplify the product.

Students spent $2\frac{4}{5}$ of their working time babysitting.

Multiply. Write the answer in simplest form.

1. $\frac{1}{2} \times 6$ _____
2. $\frac{1}{3} \times 8 =$ _____
3. $\frac{4}{5} \times 15 =$ _____
4. $9 \times \frac{1}{3} =$ _____

5. $3 \times \frac{1}{6} =$ _____
6. $4 \times \frac{2}{4} =$ _____
7. $5 \times \frac{4}{9} =$ _____
8. $4 \times \frac{5}{20} =$ _____

9. $\frac{1}{18} \times 6 =$ _____
10. $\frac{3}{14} \times 8 =$ _____
11. $\frac{1}{2} \times 5 =$ _____
12. $\frac{1}{3} \times 4 =$ _____

13. $\frac{1}{5} \times 8 =$ _____
14. $9 \times \frac{1}{4} =$ _____
15. $7 \times \frac{1}{3} =$ _____
16. $\frac{2}{4} \times 5 =$ _____

Solve.

17. At West School, 15 sixth-grade students play in the band. Of these, $\frac{1}{3}$ play the drums. How many sixth graders at West School play the drums?

_____

RETEACH: Estimating a Fraction of a Number

Lance notices a sale while shopping. About how much would he save if he bought the shoes on sale?

Remember

Round the whole number to a number that is easily divided.

Round 40 to a number that is easily divided by 3.

$40 \rightarrow $39 $40 \rightarrow $42
$\frac{1}{3} \times $39 = $13   OR $\frac{1}{3} \times $42 = $14

Lance will save $13 to $14.

Round the number.

1. $\frac{1}{2}$ of $29   2. $\frac{1}{3}$ of $55   3. $\frac{1}{6}$ of 25   4. $\frac{1}{10}$ of $48
   $\frac{1}{2}$ of _____ $\frac{1}{3}$ of _____ $\frac{1}{6}$ of _____ $\frac{1}{10}$ of _____

5. $\frac{1}{4}$ of 140   6. $\frac{1}{8}$ of 165   7. $\frac{1}{5}$ of 123   8. $\frac{1}{3}$ of 100
   $\frac{1}{4}$ of _____ $\frac{1}{8}$ of _____ $\frac{1}{5}$ of _____ $\frac{1}{3}$ of _____

Estimate.

9. $\frac{1}{3}$ of $50 _____   10. $\frac{1}{5}$ of $23 _____   11. $\frac{1}{8}$ of $75 _____

12. $\frac{1}{4}$ of $35 _____   13. $\frac{1}{10}$ of 118 _____   14. $\frac{1}{2}$ of 175 _____

15. $\frac{1}{6}$ of 235 _____   16. $\frac{1}{5}$ of 139 _____   17. $\frac{2}{3}$ of $50 _____

18. $\frac{2}{5}$ of $23 _____   19. $\frac{3}{8}$ of $75 _____   20. $\frac{3}{4}$ of $35 _____

Estimate. Write > or < for ◯.

21. $\frac{1}{5}$ of 58 ◯ 10   22. $\frac{1}{3}$ of 98 ◯ 30   23. $\frac{1}{4}$ of 115 ◯ 40

24. $\frac{1}{8}$ of 235 ◯ 30   25. $\frac{2}{5}$ of 37 ◯ 14   26. $\frac{2}{3}$ of 58 ◯ 45

27. $\frac{1}{4}$ of $3.29 ◯ $0.80   28. $\frac{1}{5}$ of $98 ◯ $25   29. $\frac{1}{8}$ of $7.78 ◯ $1

Use with pages 234–235.

RETEACH: Multiplying Mixed Numbers

Frederico had a grape-picking contract with his father. Frederico received $\frac{1}{8}$ of the grapes he picked. His dad received the remaining grapes. If Frederico picked $27\frac{1}{3}$ pounds of grapes, how many pounds of grapes did he get to keep?

Remember

To multiply a mixed number by a fraction, rename the mixed number with a fraction; then multiply.

Rename the mixed number.	Multiply the fractions.	Simplify the product.
$27\frac{1}{3} = \frac{82}{3}$	$\frac{82}{3} \times \frac{1}{8} = \frac{82}{24}$	$\frac{82}{24} = \frac{41}{12} = 3\frac{5}{12}$

Frederico got $3\frac{5}{12}$ pounds of grapes.

Multiply. Write the answer in simplest form.

1. $\frac{1}{4} \times 3\frac{1}{2} =$ _____
2. $\frac{5}{8} \times 2\frac{1}{3} =$ _____
3. $1\frac{3}{4} \times \frac{2}{3} =$ _____

4. $9\frac{1}{3} \times \frac{1}{2} =$ _____
5. $3\frac{1}{4} \times \frac{1}{2} =$ _____
6. $3\frac{3}{4} \times 7 =$ _____

7. $2\frac{2}{5} \times 5 =$ _____
8. $3\frac{1}{8} \times 4 =$ _____
9. $3 \times 6\frac{1}{2} =$ _____

10. $9 \times 9\frac{1}{2} =$ _____
11. $6\frac{3}{4} \times 14 =$ _____
12. $7\frac{1}{8} \times 12 =$ _____

Solve.

13. Frederico had $3\frac{5}{12}$ pounds of grapes. He sold them for 60 cents per pound. How much money did Frederico receive for his grapes?

   _____

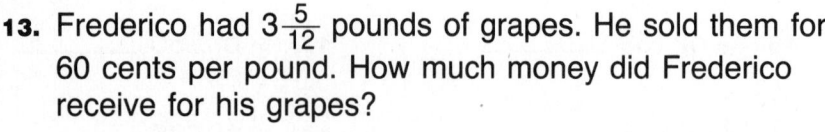

RETEACH: Dividing Whole Numbers and Fractions

A baker uses $\frac{3}{4}$ pound butter to make 1 batch of muffins. How many batches of muffins can the baker make with 6 pounds butter?

Remember

To find how many $\frac{3}{4}$ there are in 6, or what number times $\frac{3}{4}$ will produce 6, use a reciprocal. The reciprocal of $\frac{3}{4}$ is $\frac{4}{3}$. Rather than dividing by a fraction, you can multiply by the reciprocal of the fraction.

$$6 \div \frac{3}{4} \rightarrow 6 \times \frac{4}{3}$$
$$\rightarrow \frac{6}{1} \times \frac{4}{3}$$
$$\rightarrow \frac{6 \times 4}{1 \times 3} \rightarrow \frac{24}{3} = 8$$

The baker can make 8 batches of muffins.

Write the reciprocal of each number.

1. $\frac{5}{8} =$ _____
2. $\frac{3}{4} =$ _____
3. $\frac{7}{10} =$ _____
4. $2\frac{1}{5} =$ _____
5. $\frac{6}{1} =$ _____

6. $\frac{3}{2} =$ _____
7. $\frac{3}{1} =$ _____
8. $\frac{7}{1} =$ _____
9. $7 =$ _____
10. $3\frac{1}{4} =$ _____

Divide. Write the answer in simplest form.

11. $3 \div \frac{1}{4} =$ _____
12. $2 \div \frac{1}{3} =$ _____
13. $\frac{4}{5} \div \frac{1}{5} =$ _____
14. $5 \div \frac{1}{2} =$ _____

15. $4 \div \frac{1}{2} =$ _____
16. $\frac{3}{8} \div 4 =$ _____
17. $1\frac{3}{4} \div \frac{3}{7} =$ _____
18. $6 \div 1\frac{1}{2} =$ _____

Solve.

19. People's Bakery uses $\frac{1}{4}$ sack of flour for every batch of bread. How many batches of bread can they make if they have 12 sacks of flour?

_____

Use with pages 240–241.

RETEACH: Dividing Fractions and Mixed Numbers

Julie was taking piano lessons from Mrs. Wymore. If Julie must practice $6\frac{1}{4}$ hours in 5 days, how many hours per day must she practice?

Remember

To divide by 5, we can multiply by the reciprocal of 5.

Since $5 = \frac{5}{1}$, the reciprocal of $\frac{5}{1}$ is $\frac{1}{5}$.

$$6\frac{1}{4} \div 5 \rightarrow \frac{25}{4} \div 5 \rightarrow \frac{25}{4} \div \frac{5}{1}$$

$$\frac{25}{4} \times \frac{1}{5} = \frac{25}{20}$$

$$\frac{25}{20} = \frac{5}{4}$$

$$\frac{5}{4} = 1\frac{1}{4}$$

Julie must practice for $1\frac{1}{4}$ hours each day.

Divide. Simplify the answer.

1. $6\frac{3}{4} \div \frac{1}{2} =$ _____
2. $8\frac{1}{3} \div \frac{2}{3} =$ _____
3. $8\frac{3}{4} \div \frac{4}{5} =$ _____

4. $1\frac{7}{10} \div \frac{1}{10} =$ _____
5. $4\frac{1}{4} \div \frac{5}{8} =$ _____
6. $6\frac{1}{2} \div 8 =$ _____

7. $7\frac{1}{2} \div 3 =$ _____
8. $6\frac{1}{3} \div 2 =$ _____
9. $4\frac{1}{2} \div 3 =$ _____

Solve.

10. It takes Julie $1\frac{1}{2}$ minutes to play her piano recital piece. How many times can she practice the piece in 6 minutes?

_____

11. Julie practiced $14\frac{1}{2}$ hours in two weeks. She did not practice on Saturday or Sunday in either week. How many hours per day did she practice in those two weeks?

_____

RETEACH Measuring Customary Lengths

Alan is going fishing with his grandfather. Grandfather says they need fishhooks that are at least $1\frac{1}{2}$ inches long. Is this fishhook long enough?

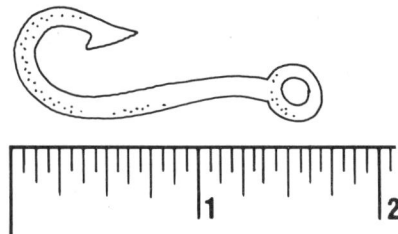

Remember

Count units to the closest given measure. The length of the fishhook is:

2 in. to the nearest in.

$1\frac{1}{2}$ in. to the nearest $\frac{1}{2}$ in.

$1\frac{3}{4}$ in. to the nearest $\frac{1}{4}$ in.

$1\frac{6}{8}$ in. to the nearest $\frac{1}{8}$ in.

$1\frac{11}{16}$ in. to the nearest $\frac{1}{16}$ in.

The fishhook measures $1\frac{11}{16}$ in. long.

Yes, it is long enough.

Choose the unit you would use to measure

1. the length of a football field. _____
2. the length of a tennis racquet. _____
3. the height of your school building. _____
4. the height of newborn baby. _____
5. the length of a school bus. _____
6. the length of a soccer field. _____
7. the length of North Dakota. _____
8. the distance from Chicago to Dallas. _____

Measure this fishing lure to the nearest

9. in.
10. $\frac{1}{2}$ in.
11. $\frac{1}{4}$ in.
12. $\frac{1}{16}$ in.

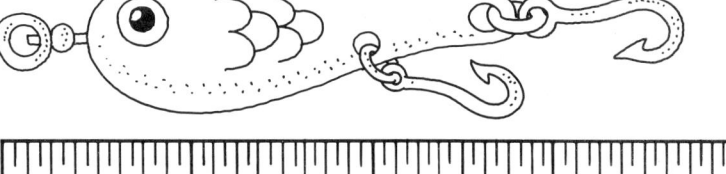

Use with pages 246–247.

RETEACH — Customary Units of Length

On a baseball diamond, the distance from home plate to first base is 90 feet. How many yards is that?

> **Remember**
>
> 12 inches (in.) = 1 foot (ft)
> 3 feet (ft) = 1 yard (yd)
> 1,760 yards (yd) = 1 mile (mi)
> 5,280 feet (ft) = 1 mile (mi)
>
> To rename a smaller unit with a larger unit, we must divide. To rename a larger unit with a smaller unit, we must multiply. Since we rename feet (smaller units) with yards (larger units), we must divide by 3.
>
> $$90 \div 3 = 30$$

The distance from home plate to first base is 30 yards.

Complete.

1. 15 yd = _____ ft
2. 4 mi = _____ yd
3. 81 ft = _____ yd
4. 5,280 yd = _____ mi
5. 6 ft 2 in. = _____ in.
6. 5 yd 2 ft = _____ ft
7. 2 mi 30 yd = _____ yd
8. $5\frac{1}{3}$ yd = _____ ft

9.   2 ft 7 in.
  + 3 ft 4 in.

10.   12 yd 1 ft
  + 6 yd 2 ft

11.   3 mi 560 yd
  + 1 mi 1200 yd

12.   80 ft 9 in.
  − 30 ft 7 in.

13.   5 yd 1 ft
  − 2 yd 2 ft

14.   7 mi 600 yd
  − 3 mi 1530 yd

Solve.

15. A football field is 100 yards long and $53\frac{1}{3}$ yards wide. Find the length and width of a football field in feet.

_____

RETEACH: Customary Units of Capacity and Weight

The zoo keeper at the Westwood Zoo told the sixth-grade class that Jumbo, the elephant, has a mass of 6 tons. What is Jumbo's mass in pounds?

Remember

8 fluid ounces (fl oz) = 1 cup (c)
2 cups (c) = 1 pint (pt)
2 pints (pt) = 1 quart (qt)
4 quarts (qt) = 1 gallon (gal)

16 ounces (oz) = 1 pound (lb)
2,000 pounds (lb) = 1 ton (T)

To rename a larger unit with a smaller unit, we must multiply. To rename tons (a larger unit) with pounds (a smaller unit), we must multiply by 2,000.

$$6 \times 2{,}000 = 12{,}000$$

Jumbo has a mass of 12,000 pounds.

Complete.

1. 10 pt = _____ c
2. 3 qt = _____ pt
3. 27 lb = _____ oz
4. 6 gal = _____ qt
5. 48 c = _____ pt
6. 22 pt = _____ qt
7. 368 oz = _____ lb
8. 6,000 lb = _____ T
9. 4 T = _____ lb
10. 36 c = _____ pt
11. 23 gal = _____ qt
12. 48 oz = _____ lb

Solve.

13. If Jumbo drinks 75 gallons of water each day, how many quarts of water does she drink each day?

_____

Use with pages 250–251.

RETEACH: Units of Time: Addition and Subtraction

In the early days of sea travel, it often took as many as 432 hours for a ship to cross the Atlantic Ocean. How many days is that?

Remember

60 seconds (s)	= 1 minute (min)
60 minutes (min)	= 1 hour (h)
24 hours (h)	= 1 day (d)
7 days (d)	= 1 week (wk)
365 days (d)	= 1 year (y)
52 weeks (wk)	= 1 year (y)
12 months (mo)	= 1 year (y)
100 years (y)	= 1 century

To rename a smaller unit with a larger unit, we must divide. To rename hours (a small unit) with days (a larger unit), we must divide by 24, the number of hours in a day.

$$432 \div 24 = 18$$

It took 18 days to cross the Atlantic.

Complete.

1. 5 min = _____ s
2. 10 h = _____ min
3. 3 h = _____ min
4. 6 d = _____ h
5. 1 y 2 mo = _____ mo
6. 4 min 25 s = 3 min _____ s
7. 3 d 10 h = 2 d _____ h
8. 2 d = _____ hr

Complete.

9. 3 min 28 s
   + 6 min 16 s

10. 10 h 2 min
    + 2 h 35 min

11. 6 h 34 min
    + 2 h 28 min

Solve.

12. When grandfather Johannas crossed the Atlantic Ocean in 1907, it took 14 days 21 hours. How many hours was that? _____

RETEACH   Temperature: Fahrenheit

The highest temperature recorded in the United States is 134°F in Death Valley, California, on July 10, 1913. The lowest temperature recorded in the United States is nearly ⁻80°F in Alaska on January 23, 1971. How many degrees are there between the highest and lowest temperatures?

Remember

Count from the highest degree to zero, and then from zero to the lowest degree. From 134 to 0 is 134. From 0 to ⁻80 is 80.

$$134 + 80 = 214$$

There are 214°F between 134°F and ⁻80°F.

How many degrees are there between

1. 80°F and 30°F _____
2. 74°F and 24°F _____
3. 64°F and 22°F _____

4. 103°F and 20°F _____
5. 79°F and ⁻15°F _____
6. 48°F and ⁻42°F _____

7. 100°F and ⁻50°F _____
8. 32°F and 0°F _____
9. 0°F and ⁻40°F _____

10. 40°F and ⁻40°F _____
11. 72°F and ⁻2°F _____
12. 2°F and ⁻23°F _____

Solve.

13. The most extreme temperatures recorded in Texas are 120°F and ⁻23°F. How many degrees are there between these temperatures? _____

14. In one day, the temperature in Fort Walton Beach, Florida, climbed from 68°F to 102°F. How many degrees did the temperature rise in that period of time? _____

Use with pages 254–255.

RETEACH   Ratios

Scott took 7 rolls of film with him to summer camp. He used 5 rolls of film. Compare the number of rolls of film Scott used to the number of rolls of film he had.

Remember

To make a correct comparison, be careful to write the numbers in the correct order. You may write 5 of 7 because you are comparing a smaller number to the total.

$\frac{5}{7}$ can be written in three different ways.

$\frac{5}{7}$ **or**   5 to 7   **or**   5:7

Scott used 5 of the 7 rolls, or $\frac{5}{7}$.

Express these ratios as fractions.

1. 2 to 3 _____   **2.** 4 to 5 _____   **3.** 5 to 2 _____   **4.** 6 to 1 _____

5. 7:8 _____   **6.** 1:2 _____   **7.** 100:1 _____   **8.** 15:100 _____

Express each ratio in two different ways.

9. 5:3   **10.** $\frac{3}{4}$ **11.** 10 to 1   **12.** $\frac{1}{100}$ **13.** 9 to 5

Solve.

14. There were 143 campers at Emile's camp. Of the campers, 73 were girls. Write a ratio that compares the number of girls to the number of boys.

RETEACH Equal Ratios

In its first year of operation, $\frac{2}{5}$ of the sixth-grade students at Gerry's basketball camp were girls. If there were 40 girls, what was the total number of sixth-grade students at the camp?

Remember

To find equivalent fractions, or equal ratios, multiply both the numerator and the denominator by the same number.

$$\frac{2}{5} \begin{array}{c} \leftarrow girls \rightarrow \\ \leftarrow total \rightarrow \end{array} \frac{40}{n}$$

$$\frac{2 \times 20}{5 \times 20} = \frac{40}{100}$$

Also note that when cross products are equal, the ratios are equal.

$$\frac{2}{5} = \frac{40}{100} \qquad \begin{array}{c} 2 \times 100 = 5 \times 40 \\ (200) \qquad (200) \end{array}$$

There are 100 sixth-grade students at the camp.

Find the missing number.

1. $\frac{2}{3} = \frac{4}{\underline{}}$
2. $\frac{5}{7} = \frac{\underline{}}{14}$
3. $\frac{\underline{}}{5} = \frac{2}{10}$
4. $\frac{3}{\underline{}} = \frac{75}{100}$
5. $\frac{15}{10} = \frac{\underline{}}{2}$
6. $\frac{7}{5} = \frac{21}{\underline{}}$
7. $\frac{\underline{}}{8} = \frac{4}{32}$
8. $\frac{50}{\underline{}} = \frac{2}{1}$

Solve.

9. During the second year of Gerry's basketball camp, the ratio of girls to boys was $\frac{3}{4}$. If there were 80 boys at the camp, how many girls were there?

_____

Use with pages 268–269.

79

RETEACH   Proportions

At a summer computer camp, there are 3 instructors for every 20 campers. How many instructors will be needed if there are 80 campers?

Remember

Find a proportion that states that two ratios are equal.

$$\frac{\text{instructors}}{\text{campers}} = \frac{\text{instructors}}{\text{campers}}$$

$$\frac{3}{20} = \frac{n}{80}$$

Multiply to find the cross product. $3 \times 80 = 20 \times n$
Solve the equation. $240 = 20 \times n$
$240 \div 20 = n$
$12 = n$

The camp will need 12 instructors.

Find the proportion.

1. $\frac{3}{5} = \frac{9}{}$
2. $\frac{1}{3} = \frac{}{12}$
3. $\frac{}{5} = \frac{12}{20}$
4. $\frac{6}{} = \frac{18}{3}$
5. $\frac{}{2} = \frac{25}{50}$
6. $\frac{3}{} = \frac{30}{40}$
7. $\frac{9}{5} = \frac{}{15}$
8. $\frac{75}{100} = \frac{3}{}$

Solve.

9. A summer computer camp has 4 computers for each 10 campers. How many campers will 20 computers accommodate?

10. At the computer camp, 5 campers can run 15 programs on their computers in an hour. How many programs could 9 campers run in one hour?

11. One of the computers at camp can compute 60 difficult calculations in 3 seconds. How many of the same calculations can the computer compute in 5 seconds?

RETEACH Scale Drawings

The scale of the gears at the right is given as 1 in. = 6 in. What is the actual diameter of gear A?

Remember

The diameter of gear A in the drawing is 2 in.

Write a proportion and solve.

Length in drawing → $\frac{1}{6} = \frac{2}{d}$
Actual length →

$1 \times d = 2 \times 6$
$d = 12$

The actual diameter of gear A is 12 in.

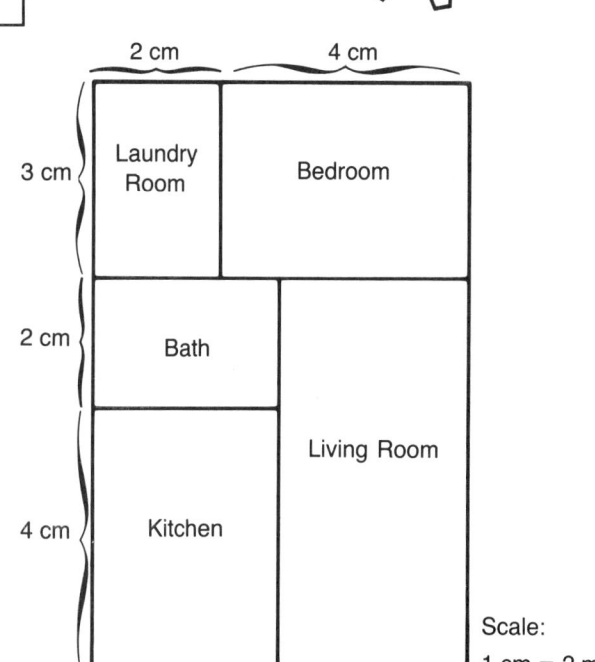

Scale: 1 cm = 2 m

Use the scale drawing at the right to answer the following questions.

1. How wide is the kitchen?

2. How long is the kitchen?

3. How wide is the laundry room?

4. How wide is the bedroom?

5. How long is the house?

6. How wide is the house?

7. How long is the living room?

8. How long is the bathroom?

Solve.

9. The scale on a house plan is 1 in. = 4 ft. How long is the actual house if the scale drawing of the house is 12 in. long and 9 in. wide?

Use with pages 276–277.

RETEACH Percent

Puffed wheat has 15 parts of protein for each 100 parts of cereal. Write this ratio as a percent.

Remember

Since the ratio has a denominator of 100, we can change the ratio to a percent by using the numerator and the percent (%) symbol.

$$\frac{15}{100} = 15\%$$

Puffed wheat cereal is 15% protein.

Write the percent for each ratio.

1. 61 to 100 _____
2. 13 to 100 _____
3. 4 to 100 _____
4. 12 to 100 _____

5. 3:100 _____
6. 15:100 _____
7. 88:100 _____
8. 1:100 _____

Write a percent for each decimal.

9. 0.75 _____
10. 0.81 _____
11. 0.69 _____
12. 0.11 _____

13. 0.1 _____
14. 0.3 _____
15. 0.07 _____
16. 0.05 _____

Solve.

17. Puffed wheat is 0.73 starch. Write 0.73 as a percent.

18. John only eats puffed wheat for breakfast about 0.02 of the time. Write a percent to show how many times John eats puffed wheat for breakfast. _____

RETEACH: Fractions and Decimals for Percents

The *wheat kernel,* sometimes called the *wheat berry,* is the seed from which the wheat plant grows. Each tiny seed contains three distinct parts that are separated during the milling process in order to produce flour. The endosperm makes up 83% of the kernel weight. Write 83% as a decimal.

Remember

To rename a percent with a decimal, think of it as a fraction that has a denominator of 100. Then write it as a decimal in terms of hundredths.

$$83\% = \frac{83}{100} = 0.83$$

So, 83% is 0.83.

Write each percent as a fraction in simplest form.

1. 40% _____   2. 31% _____   3. 57% _____   4. 90% _____   5. 7% _____

6. 12% _____   7. 75% _____   8. 30% _____   9. 3% _____   10. 98% _____

Write each percent as a decimal.

11. 81% _____   12. 73% _____   13. 42% _____   14. 25% _____   15. 3% _____

16. 90% _____   17. 4% _____   18. 50% _____   19. 33% _____   20. 15% _____

Solve.

21. A kernel of wheat is 14.5% bran and 2.5% germ. Write these percents as fractions and decimals.

_____

Use with pages 282–283.

RETEACH Percents for Fractions

The West School has enough computers for 12 members of each class. The sixth-grade class at West School has 25 students. What percent of the sixth-grade students can use the computers at any one time?

Remember

First write the ratio as a fraction.

$$\frac{\text{Students on computers}}{\text{Total sixth-grade students}} = \frac{12}{25}$$

Write an equivalent fraction that has a denominator of 100. Then write the fraction as a percent.

$$\frac{12}{25} = \frac{12 \times 4}{25 \times 4} = \frac{48}{100} = 48\%$$

48% of the students can use the computers at any one time.

Write a percent for each fraction.

1. $\frac{3}{25}$ _____
2. $\frac{7}{10}$ _____
3. $\frac{5}{20}$ _____
4. $\frac{8}{50}$ _____
5. $\frac{8}{32}$ _____

6. $\frac{8}{40}$ _____
7. $\frac{24}{40}$ _____
8. $\frac{48}{80}$ _____
9. $\frac{18}{50}$ _____
10. $\frac{30}{75}$ _____

11. $\frac{3}{5}$ _____
12. $\frac{2}{25}$ _____
13. $\frac{9}{20}$ _____
14. $\frac{6}{50}$ _____
15. $\frac{27}{30}$ _____

Solve.

16. The Mathematics Club of South School bought computers for the sixth-grade room. Now they have 14 computers for the 25 students in the room. Write the ratio, computers to students, as a fraction and as a percent.

_____

RETEACH — Percent of a Number

A poll showed that 40% of the 30 students in the sixth grade had home computers. How many sixth-grade students have computers in their homes?

Remember

To find a percent of a number, rename the percent with a decimal or a fraction. Then do the calculation.

Rename the percent with a fraction.

$$40\% = \frac{40}{100} = \frac{2}{5}$$

Rename the percent with a decimal.

$$40\% = \frac{40}{100} = 0.40$$

Calculate.

$$\frac{2}{5} \times 30 = 12 \qquad 0.40 \times 30 = 12$$

12 of the sixth-grade students have computers in their homes.

Find the percent of each number.

1. 6% of 30 _____
2. 9% of 16 _____
3. 12% of 50 _____
4. 60% of 250 _____
5. 58% of 310 _____
6. 95% of 150 _____

Solve.

7. The sixth-grade class had a birthday party for their teacher. They ate 10 pounds of nut bread. If the nut bread is 30% pecans, how many pounds of pecans did the class eat? _____

8. On Valentine Day, the sixth-grade class made cards to give to people in the community. They made 80 cards in all and gave 40% of them to the hospital. How many valentines did they give to the hospital? _____

Use with pages 286–287.

RETEACH: Finding Percents

Of the 50 states of the United States, 24 lie west of the Mississippi River. What percent of the states are west of the Mississippi?

Remember

Answer this question:
What percent of 50 is 24?

Write an equation. $n \times 50 = 24$
$$n = 24 \div 50$$
$$n = 0.48 = 48\%$$

Of the 50 states, 48% are west of the Mississippi.

Find the percent.

1. What percent of 36 is 9? _____
2. What percent of 35 is 7? _____
3. What percent of 250 is 62.5? _____
4. What percent of 50 is 5? _____
5. What percent of 84 is 21? _____
6. What percent of 95 is 19? _____
7. What percent of 108 is 54? _____
8. What percent of 16 is 4? _____
9. 17 is what percent of 68? _____
10. 13 is what percent of 65? _____
11. 312.5 is what percent of 250? _____
12. 128.25 is what percent of 95? _____

Solve.

13. The Atlantic Ocean borders 14 of the 50 states in the United States. What percent of the 50 states border the Atlantic Ocean? _____

14. Of the 14 states that border the Atlantic Ocean, 4 have names that start with *N*. They are New Hampshire, New York, New Jersey, and North Carolina. There are 4 other states that start with *N*. What percent of the 8 states that start with *N* border the Atlantic? _____

RETEACH: Finding the Total Number

Mr. Schwieso raises corn in Iowa. After harvesting 45 acres of corn, he has completed 30% of his harvest. How many acres of corn does he have?

Remember

Answer this question:
30% of what number is 45?

Write an equation. $30\% \times n = 45$
Write 30% as a decimal. $0.30 \times n = 45$
Solve. $n = 45 \div 0.30$
$n = 150$

Mr. Schwieso has 150 acres of corn.

Find the number.

1. 15% of what number is 12? _____
2. 50% of what number is 13? _____
3. 20% of what number is 9? _____
4. 40% of what number is 150? _____
5. 90% of what number is 45? _____
6. 75% of what number is 51? _____
7. 15 is 30% of what number? _____
8. 45 is 50% of what number? _____
9. 68 is 25% of what number? _____
10. 4 is 8% of what number? _____

Solve.

11. Mr. Schwieso has 32 bushels of Silver Queen corn. That amount is 80% of his total corn crop. How many bushels of corn does he have? _____

12. Annie Schwieso shucks corn to help her father. In one day, Annie was able to shuck 20 bushels of the Bread-and-Butter corn that her father wanted to sell at his roadside stand. This was 25% of the total. How many bushels of corn did they have? _____

Use with pages 290–291.

RETEACH Basic Vocabulary of Geometry

A utility pole ($\overline{EF}$ in the figure at the right) is braced by two cables, $\overline{BA}$ and $\overline{CD}$. Name the angle that is formed by the pole $\overline{EF}$ and the cable $\overline{BA}$.

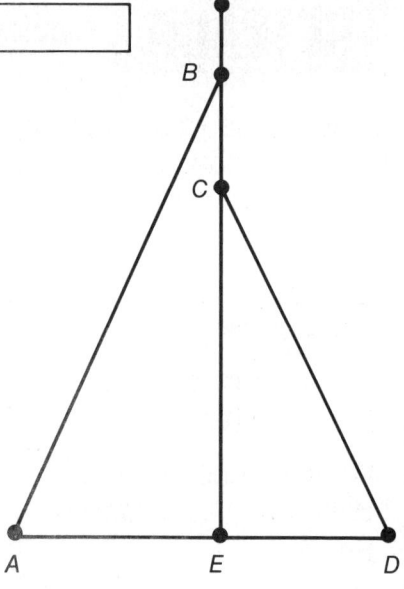

Remember
To name an angle, you must use three letters. The middle letter is always the vertex. There are two ways we could name the angle formed by $\overline{AB}$ and $\overline{EF}$: either ∠ABC or ∠ABE.

The angle formed by pole $\overline{EF}$ and cable $\overline{BA}$ can be called either ∠ABC or ∠ABE.

Use the figure at the right to name each.

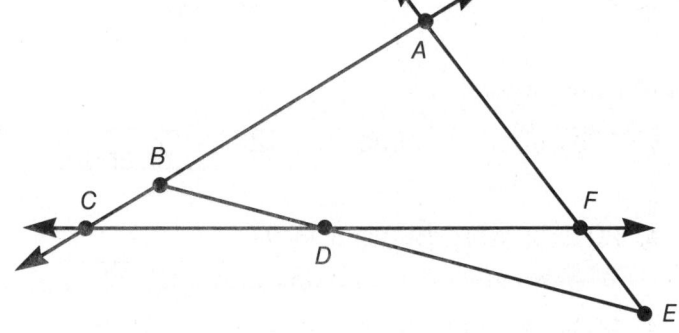

1. $\overrightarrow{EA}$ _____
2. $\overline{BE}$ _____
3. ∠AFD _____
4. $\overline{EF}$ _____
5. D _____
6. ∠AEB _____
7. $\overleftrightarrow{CF}$ _____
8. $\overrightarrow{DC}$ _____
9. F _____
10. $\overleftrightarrow{CA}$ _____

11. The figure at the right is a plan of the runways at Skyhappy Airport. Write how many

   angles. _____
   rays. _____
   points. _____
   lines. _____
   line segments. _____

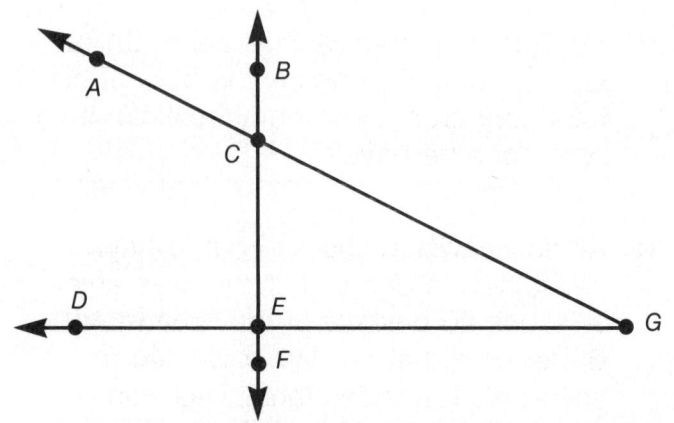

Use with pages 306–307.

RETEACH   Angles

Timothy is designing a robot that can do yard work. He wants to label the angles in the drawing as *right, obtuse,* or *acute.* How many right angles are there? how many acute angles? how many obtuse angles?

Remember

When two rays or line segments meet at a 90° angle, they form a right angle.

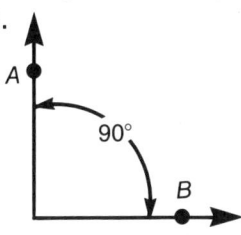

An angle that measures more than 90° but less than 180° is an obtuse angle.

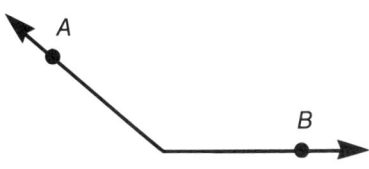

An angle that is less than 90° is an acute angle.

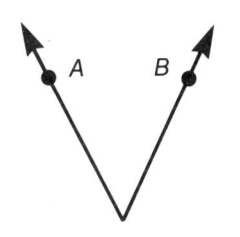

Timothy's robot drawing has two right angles, four obtuse angles, and six acute angles.

Timothy is ready to program his lawn robot to cut and trim his front lawn. Use a protractor and the diagram of his lawn at the right to answer the questions below.

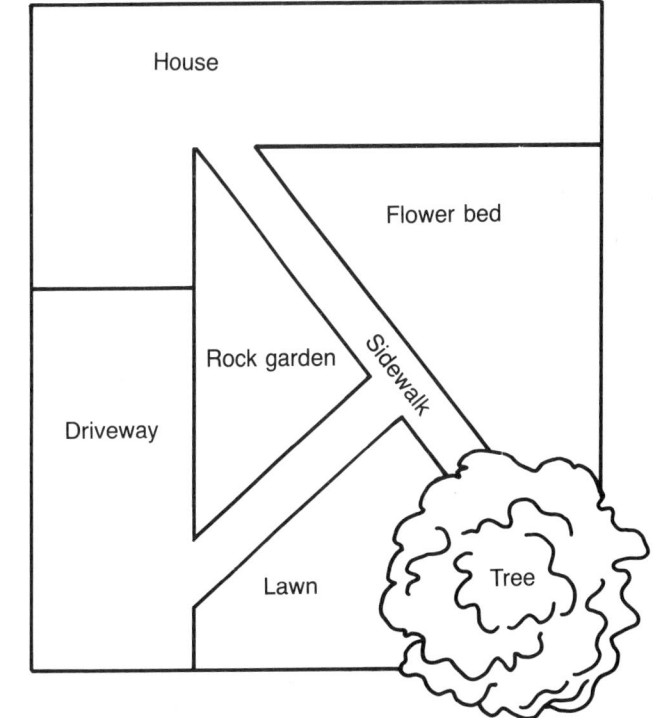

1. How many right angles are there?

2. How many obtuse angles are there?

3. How many degrees in one right angle?

4. How many degrees were formed where the sidewalks meet the lawn? _____

5. How many degrees were formed where the sidewalks meet the rock garden?

Use with pages 308–309.

RETEACH Perpendicular and Parallel Lines

Look at the figure at the right. How many pairs of parallel lines are there? What are their names? Which lines are perpendicular to each other? Which lines that are not perpendicular intersect?

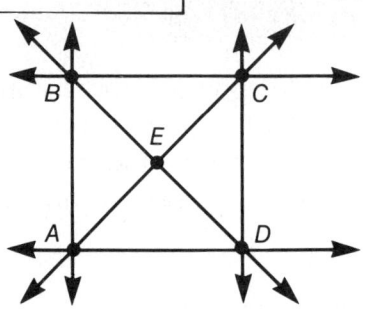

Remember

Lines that never cross are **parallel lines**.

Lines that cross are **intersecting lines**.

Lines that intersect at right angles are **perpendicular lines**.

There are two pairs of parallel lines. $\overleftrightarrow{BC} \| \overleftrightarrow{AD}$, $\overleftrightarrow{BA} \| \overleftrightarrow{CD}$; $\overleftrightarrow{AB} \perp \overleftrightarrow{BC}$ and $\overleftrightarrow{AD}$, $\overleftrightarrow{CD} \perp \overleftrightarrow{AD}$ and $\overleftrightarrow{BC}$; $\overleftrightarrow{BD}$ intersects $\overleftrightarrow{AC}$, $\overleftrightarrow{AD}$, and $\overleftrightarrow{BC}$; $\overleftrightarrow{AC}$ intersects $\overleftrightarrow{BD}$, $\overleftrightarrow{BC}$, and $\overleftrightarrow{AD}$.

Write *parallel, perpendicular,* or *intersecting* to describe each pair of lines.

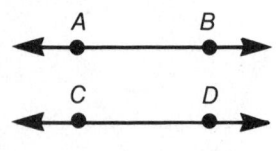

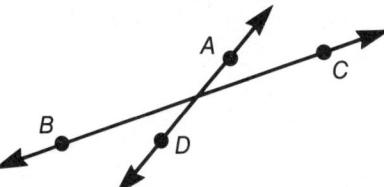

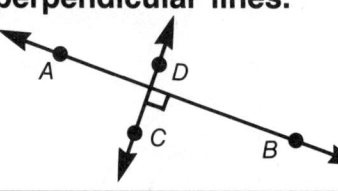

1. _____

2. _____

3. _____

Use the figure at right to solve.

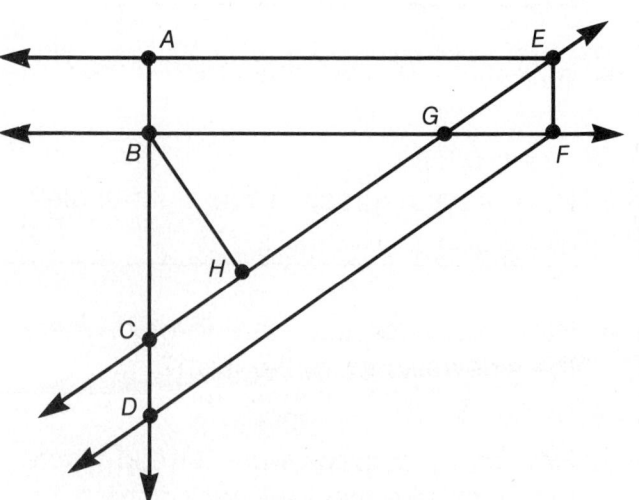

4. Name the parallel lines.
_____

5. Name 3 intersecting lines.
_____

6. Name 2 perpendicular lines.
_____

7. Name the rays.
_____

8. Name the right angles.
_____

RETEACH Triangles

Look at the figure at the right. Name the equilateral triangle. Name the isosceles triangle. How many scalene triangles are there? How many obtuse scalene triangles are there? how many right triangles?

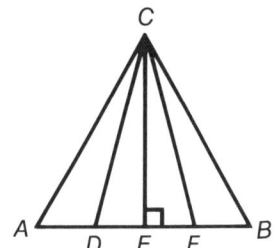

Remember

equilateral triangles have three sides of equal length.

isosceles triangles have two sides of equal length.

scalene triangles have no sides of equal length.

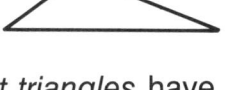

acute triangles have three acute angles (less than 90°)

obtuse triangles have one obtuse angle (more than 90°)

right triangles have one right angle (exactly 90°)

Triangle *ABC* is equilateral. Triangle *DCF* is isosceles. There are six scalene triangles. There are two obtuse triangles. There are four right triangles.

Write *equilateral*, *isosceles*, or *scalene* to describe the triangles in Exercise 1–6. Refer to the figure at the top of the page.

1. Triangle ABC _____
2. Triangle ACD _____
3. Triangle DCF _____

4. Triangle ACE _____
5. Triangle BCA _____
6. Triangle FDC _____

Write *acute*, *right*, or *obtuse* to describe the triangles in Exercise 7–12. Refer to the figure at the top of the page.

7. Triangle ADC _____
8. Triangle DCF _____
9. Triangle AEC _____

10. Triangle CEB _____
11. Triangle DEC _____
12. Triangle ABC _____

Use with pages 314–315.

RETEACH: Quadrilaterals

Look at the drawing at the right. Name the different quadrilaterals.

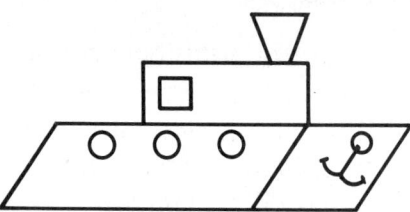

Remember

There are different kinds of quadrilaterals.

a parallelogram has two pairs of parallel sides that are of equal length.

a rhombus is a parallelogram with four equal sides.

a rectangle is a parallelogram with four right angles.

a square is a rectangle with four equal sides.

a trapezoid has only one pair of parallel sides.

The different quadrilaterals in the drawing are a trapezoid, a rhombus, a square, a rectangle, and a parallelogram.

Write the name of the shape.

1. Figure A _____
2. Figure B _____
3. Figure C _____
4. Figure D _____
5. Figure E _____
6. Figure F _____

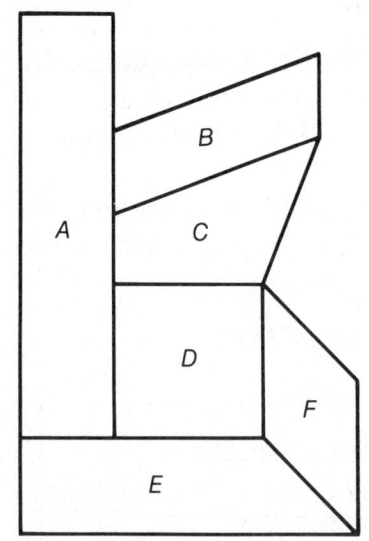

Solve.

7. Gertie Squiggle is holding two shapes. Both shapes have four sides. The first shape has two parallel sides, a 60° angle, a 70° angle, a 120° angle, and a 110° angle. The other shape has four equal sides and at least one right angle. What are the two shapes?

_____

Use with pages 318–319.

RETEACH Other Polygons

Look at the figures at the right. Which figure is an octagon? Which figure is a hexagon? Which figure is a pentagon? Name the polygon that is regular. How many sides does it have? What is the sum of the angles in the pentagon?

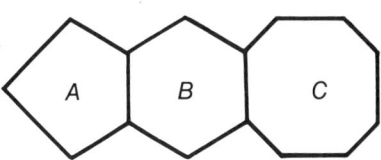

Remember

You can tell different polygons by the number of sides they have.

A *triangle* has three sides. (*Tri* is a prefix meaning "three.")

A *quadrilateral* has four sides. (*Quad* is a prefix meaning "four.")

A *pentagon* has five sides. (*Penta* is a prefix meaning "five.")

A *hexagon* has six sides. (*Hex* is a prefix meaning "six.")

An *octagon* has eight sides. (*Oct* is a prefix meaning "eight.")

A *decagon* has ten sides. (*Deca* is a prefix meaning "ten.")

A polygon is *regular* when all its sides are equal and all its angles are equal.

You can find the sum of the angles in a polygon by dividing the polygon into triangles and multiplying the number of triangles by 180°. Draw all the diagonals from one vertex.

Figure A is a pentagon. Figure B is a hexagon. Figure C is a octagon. The sum of the angles in the pentagon is 540°. The hexagon is regular. It has six sides.

Name the polygon. Write *regular* if the sides and angles are equal; write *irregular* if they are not. Find the sum of the angles.

Figure A: _____

Figure B: _____

Figure C: _____

Figure D: _____

Figure E: _____

Figure F: _____

Use with pages 320–321.

RETEACH: Congruence and Symmetry

Look at the triangles at the right. Are they congruent? Are they symmetrical?

Remember

Figures that are *exactly* the same size and have *exactly* the same shape are **congruent.** Any figure that can be divided in the middle and be the same on both sides is **symmetrical.** The dividing line is called **a line of symmetry.**

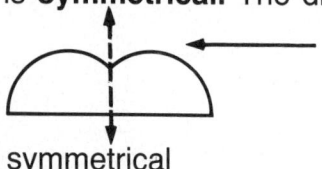 line of symmetry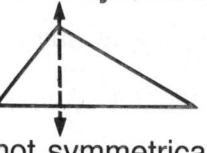

symmetrical   not symmetrical

The triangles are congruent and symmetrical.

Trace every pair of figures, write *congruent* or *not congruent*.
Draw all possible lines of symmetry.

1.

2.

3.

4.

5.

6.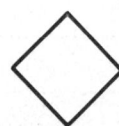

7.

8.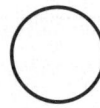

RETEACH Similar Figures

At a hobby store, you can buy models of air craft, automobiles, and ships. Each model is the same shape as the real thing but is much smaller. Figures that have the same shape but are not the same size are **similar.** Are these two figures similar?

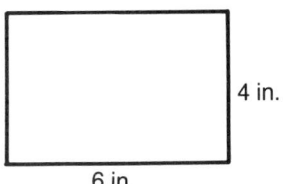

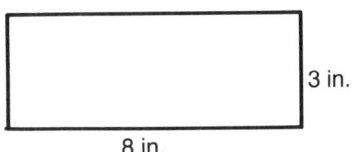

Remember

If figures are similar, they have the same shape. That is, their angles are congruent and the ratios of corresponding sides are equal.

$\frac{4}{3} \neq \frac{6}{8}$; The figures are not similar.

Write the letter of the figure that is similar to each figure in Exercises 1–4.

1. 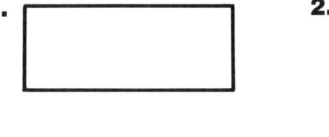   2.    A.    B.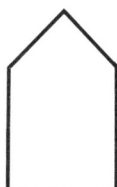

3.    4. 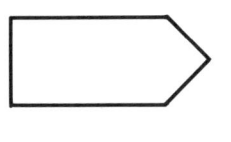   C.    D.

Given the two similar figures at the right, find the measure of the following parts.

5. ∠H = _____   6. ∠G = _____

7. ∠B = _____   8. $\overline{EF}$ = _____

9. $\overline{FG}$ = _____   10. ∠F = _____

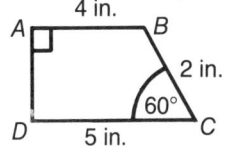

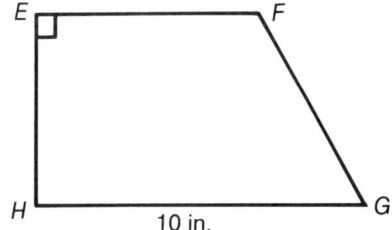

Use with pages 324–325.

95

RETEACH Circles

For the circle at the right, draw a chord, a radius, a diameter, and a central angle.

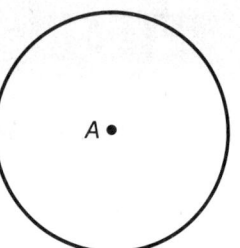

Remember

A **chord** is a line segment that has endpoints on the same circle.

A **radius** is a line segment that connects the center with any point on the circle.

A **diameter** is a chord that passes through the center of a circle; it is *always* twice the length of the radius.

A **central angle** is an angle that has its vertex at the center of the circle.

For the circle A, segment BC is a chord.
                      segment AD is a radius.
                      segment EF is a diameter.
                      angle EAD is a central angle.

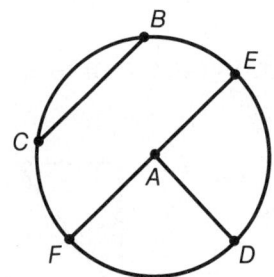

Use the circle to complete the exercises.

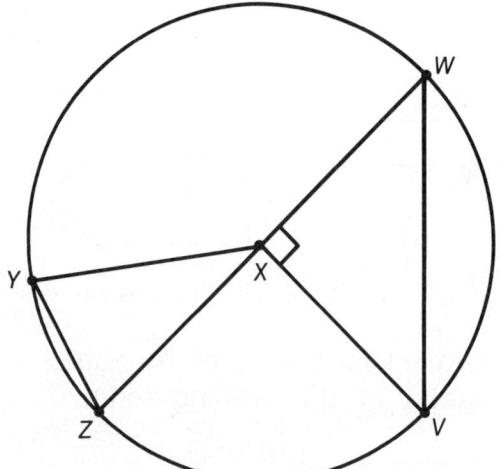

1. Name the circle.

   _____

2. Name an acute central angle.

   _____

3. Name an obtuse central angle.

   _____

4. Name a right central angle.

   _____

5. Name the center.

   _____

6. Name two chords.

   _____

7. Name three radii.

   _____

8. Name the diameter.

   _____

Use with pages 326–327.

RETEACH: Translations, Rotations, and Reflections

Look at the two figures at the right. Do they show a translation, a reflection, or a rotation?

Remember

A **translation**	A **rotation**	A **reflection**
		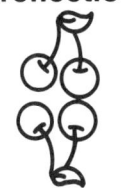
is a movement of a figure along a straight line.	is a movement of a figure along a curved line.	is a mirror image of a figure

The two figures show a reflection.

Write *translation*, *rotation*, or *reflection* to describe the relationship of Figure A to Figure B.

1.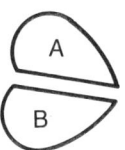

2.

3.

4.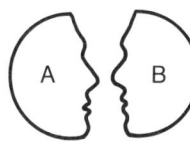

5.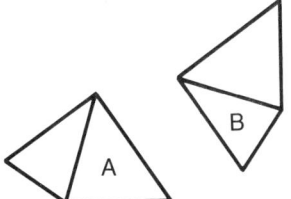

6.

Use with pages 328–329.

RETEACH Perimeter of Polygons

Tom has a paper route. This is the route he walks each day. What is the perimeter of this figure?

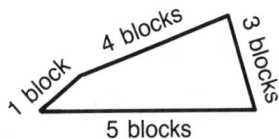

Remember

Perimeter is the distance around things. How far does Tom walk to get around this figure? He walks
5 + 3 + 4 + 1 blocks

Tom walks 13 blocks on his paper route.

Find the perimeter of each polygon.

1.

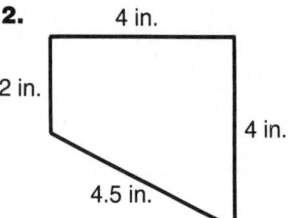

5 m, 12 m, 13 m

_____

2.

4 in., 2 in., 4 in., 4.5 in.

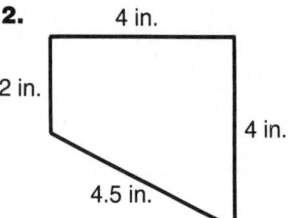

_____

3.

6 cm, 3 cm

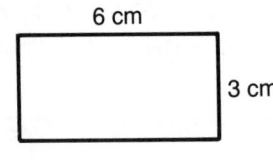

_____

4.

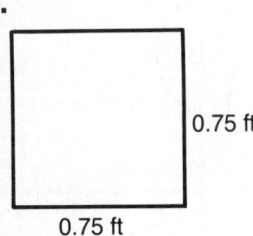

0.75 ft, 0.75 ft

_____

5.

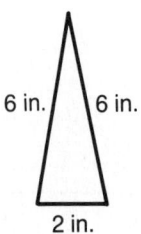

6 in., 6 in., 2 in.

_____

6.

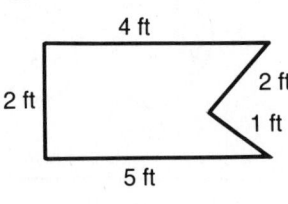

4 ft, 2 ft, 2 ft, 1 ft, 5 ft

_____

Solve.

7. A baseball diamond is really a square. The distance from home plate to first base is 90 feet. How far does a batter run when she hits a home run? _____

RETEACH: Circumference

Fido has a circular run in the backyard. The radius is 14 feet long. How far does Fido run if he runs once around the circle?

Remember

Diameter is 2 times the radius, or $2 \times 14 = 28$.

So, $C = \pi \times d \qquad \pi = \frac{22}{7}$ or 3.14

$\approx \frac{22}{7} \times 28$

$\approx \frac{22}{\underset{1}{7}} \times \overset{4}{28}$

≈ 88

Fido runs approximately 88 feet.

Find the circumference. Use $\frac{22}{7}$ for π.

1. 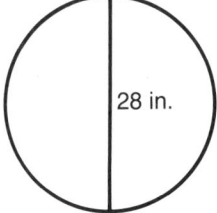 28 in.

2. 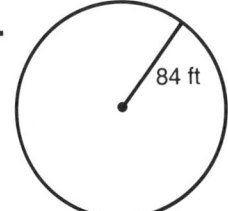 84 ft

3. 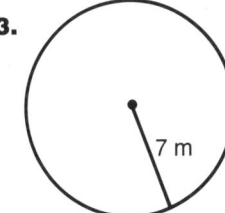 7 m

4. 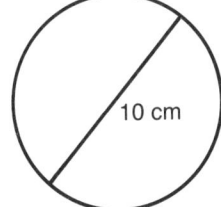 10 cm

Find the circumference. Use 3.14 for π. Round to the nearest hundredth.

5. 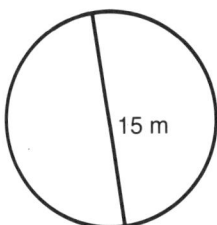 15 m

6. 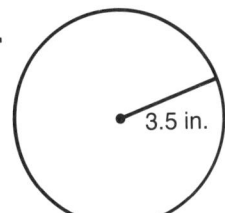 3.5 in.

7. 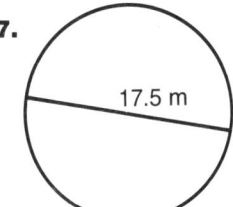 17.5 m

8. 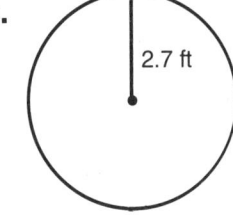 2.7 ft

RETEACH: Area of Rectangles and Squares

The playing surface of a football field is covered with artificial turf. The playing surface is 100 yards long and 50 yards wide. How many square yards of turf are needed to cover the field?

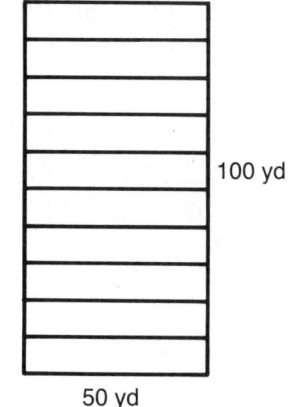
100 yd
50 yd

Remember

A football field has the shape of a rectangle. The area of a rectangle is length times width.

$$A = l \times w$$
$$A = 100 \times 50$$
$$A = 5,000$$

5,000 square yards of turf are needed.

Find the area of each figure.

1.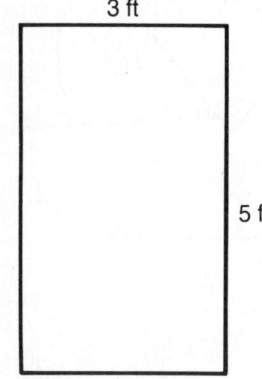
3 ft
5 ft

2.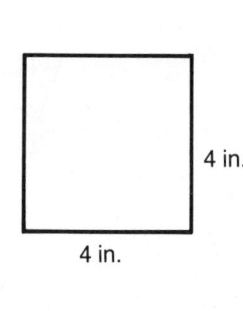
4 in.
4 in.

3.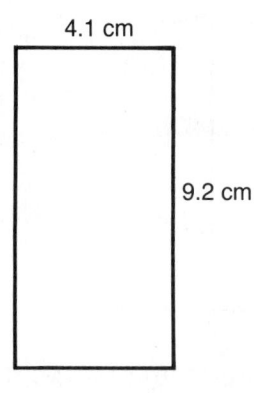
4.1 cm
9.2 cm

Find the area.

4. $l = 12$ in., $w = 7$ in. _____

5. $l = 16$ ft, $w = 3$ ft _____

6. $l = 7.5$ yd, $w = 3.5$ yd _____

7. $l = 8$ m, $w = 8$ m _____

8. $l = 49$ cm, $w = 25$ cm _____

9. $l = 1.5$ mi, $w = 1.25$ mi _____

10. $l = 15$ mm, $w = 8$ mm _____

11. $l = 27$ m, $w = 20$ m _____

12. $l = 32$ m, $w = 10$ m _____

13. $l = 12$ m, $w = 13$ m _____

Use with pages 348–349.

RETEACH — Areas of Parallelograms and Triangles

Mr. Swenson's house is located between two parallel streets. It is shaped like a parallelogram. The house is 165 ft long and the streets are 85 ft apart. The short side of the house is 90 ft. Find the area of the house.

Remember

The **height** is the distance between the two bases. The area of a parallelogram is base × height. The base here is 165 ft. The height is the distance between the streets, which is 85 ft, not 90 ft.

$$A = b \times h$$
$$A = 165 \times 85$$
$$A = 14{,}025$$

The area of the house is 14,025 ft². The area of a triangle is $A = \frac{1}{2} b \times h$.

Find the area of each figure.

1.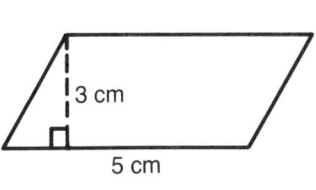

2.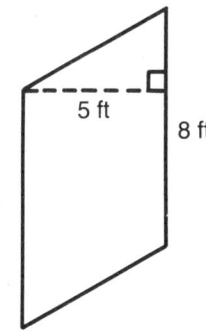

3.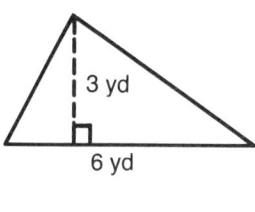

4.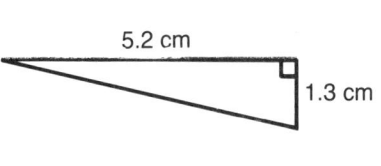

5.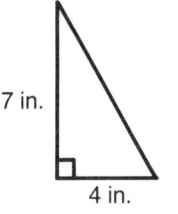

6.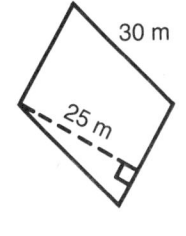

Find the area of each parallelogram.

7. $b = 5$ m, $h = 12$ m _____

8. $b = 3.2$ in., $h = 1.3$ in. _____

9. $b = 150$ ft, $h = 375$ ft _____

10. $b = 4$ m, $h = 14$ m _____

Use with pages 352–353.

RETEACH — Area of a Circle

Big Ben is the clock in the tower of the Houses of Parliament in London. Big Ben's minute hand is 14 ft long. Find the area of the circle that the minute hand covers. Use $\pi = \frac{22}{7}$.

Remember

Area $= \pi \times r^2$

$A \approx \frac{22}{7} \times 14^2$

$A \approx \frac{22}{7} \times \overset{2}{\cancel{14}} \times 14$
$\phantom{A \approx \frac{22}{7}}_{1}$

$A \approx 22 \times 28$

$A \approx 616 \text{ ft}^2$

The area of the circle is about 616 ft².

Find the area. Use $\frac{22}{7}$ for π. Round to the nearest hundredth.

1. 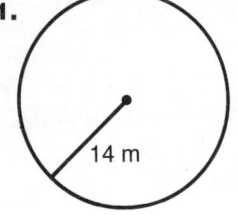   2. 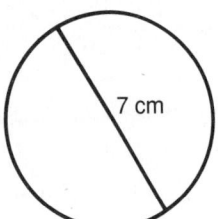   3. 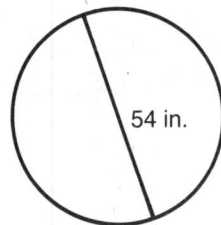   4.

_____ _____ _____ _____

5. 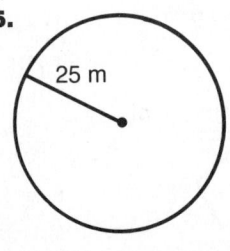   6. 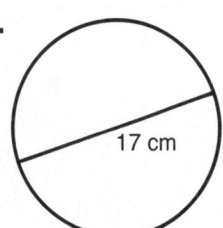   7. 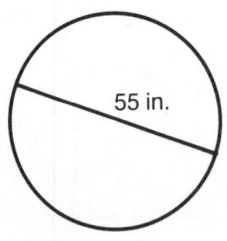   8.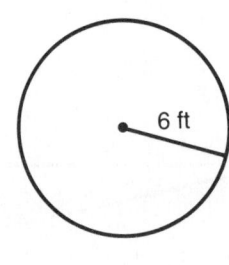

_____ _____ _____ _____

Find the area. Use $\frac{22}{7}$ for π.

9. $r = 3.14$ m _____ 10. $r = 8.4$ cm _____ 11. $r = 9.1$ cm _____

RETEACH: Solid Figures

The names of some solid shapes are *pyramid, prism, sphere, cone,* and *cylinder.* Name the shape of the building pictured.

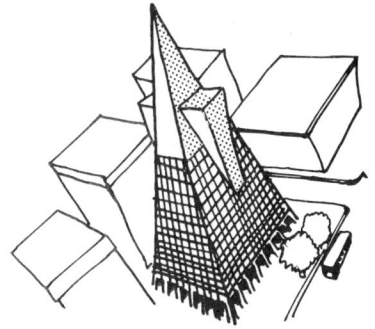

Remember

The faces of a pyramid meet at a point called the **vertex.**

It is a pyramid.

Copy and complete the chart.

	Solid figure	Name of solid figure	Number of faces	Number of edges	Number of vertices
1.		triangular pyramid	____	____	4
2.		____	6	____	8
3.		triangular prism	____	9	____
4.		____	5	8	

Write an example of each of these solid figures.

5. sphere _____   6. cone _____

7. triangular prism _____   8. rectangular prism _____

Use with pages 358–359.

RETEACH Surface Area

Airlines suggest that suitcases not be larger than a certain size. One airline allows suitcases with a length of 30 inches, a width of 20 inches, and a depth of 12 inches. Find the surface area of such a suitcase.

Remember

The **surface area** of a rectangular prism is found by adding the area of each face. Use a diagram. If the top, back, front, and ends of the suitcase were folded down, we would get a figure like this.
Area = length × width.

Face	Length	Width	Area
Top	30	12	360 in.²
Back	30	20	600 in.²
Bottom	30	12	360 in.²
Front	30	20	600 in.²
Right end	20	12	240 in.²
Left end	20	12	240 in.²
Surface area			2,400 in.²

The surface of this bag is 2,400 in.².

Find the surface area of each figure. The area of a cylinder is the area of its bases plus the area of its curved side: $2(\pi r^2) + (2\pi r \times w)$. Use $\frac{22}{7}$ for π.

1.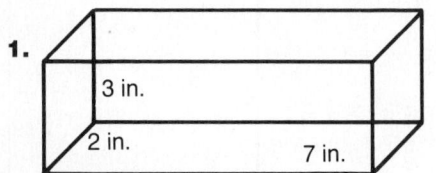

2.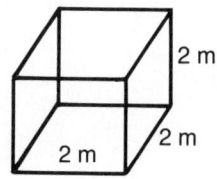

3.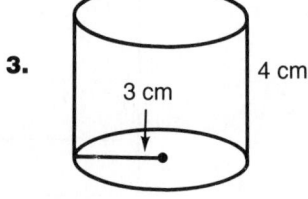

4.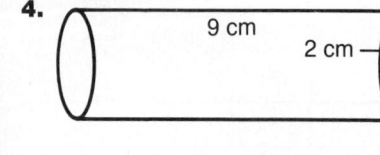

5.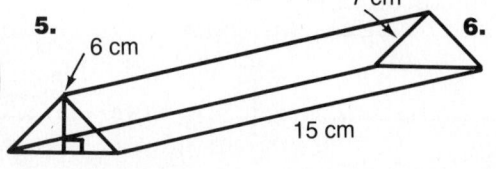

6.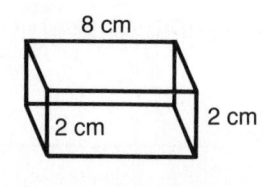

RETEACH   Volume

John has been saving pennies in a shoe box. The shoe box is 14 inches long, 7 inches wide, and 5 inches high. What is the volume of his shoe box?

Remember
Use the correct units.

$V = l \times w \times h$
$V = 14 \times 7 \times 5$
$V = 490$ cubic inches, or 490 in.^{3}.

The volume of the box is 490 in.^{3}.

Find the volume.

1.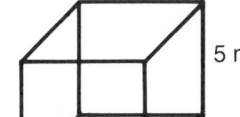

2.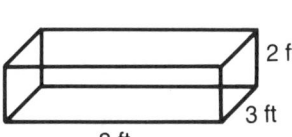

3.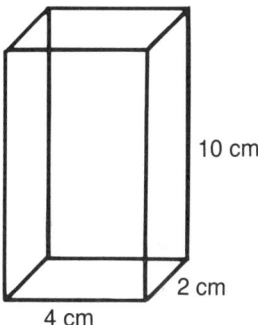

_____   _____   _____

4. $l = 10$ cm, $w = 7$ cm, $h = 5$ cm _____

5. $l = 6$ mi, $w = 4$ mi, $h = 3$ mi _____

6. $l = 7.5$ in., $w = 5.5$ in., $h = 2.5$ in. _____

Solve.

7. Mr. Ewen stores wheat in a bin shaped like a rectangular prism. The bin is 14 ft long, 12 ft wide, and 18 ft high. Find the volume of the bin. _____

Use with pages 362–363.

RETEACH: Recording Data on a Table

Mrs. Garcia asked her 30 sixth-grade students to name their favorite foods. Mrs. Garcia asked George to tally the results. After George made a tally of the information, he organized it in a table.

> **Remember**
> The table has one column for each type of data. One column is for food, and one column is for votes. The table also has a title.

After we have the data from the survey arranged in a table, we are ready to answer questions about the data.

Solve. Use the FAVORITE FOOD table.

1. What is the most favorite food? _____

2. What is the least favorite food? _____

3. Name the three most favorite foods.

   _____

4. What food is in the middle? _____

5. Is any one food liked by over half the class? _____

6. What foods are tied as favorites?

   _____

7. Did all 30 members of the class vote? _____

8. If the class had a party and they could only buy one type of food, what would they serve? _____

9. Choose one of these projects. For each one you will collect the data yourself. After you have collected the data, tally it. Then put it in a table.

Project 1: Interview the members of your family to find out their favorite foods. Set up a table that shows this information.

Project 2: Interview 6 of your classmates to find out their favorite foods. Set up a table that shows this information.

TALLY

Food	Tally
Pizza	ʇʜʅ II
Spaghetti	ʇʜʅ I
Frankfurters	ʇʜʅ
Macaroni & cheese	ʇʜʅ I
Steak	I
Hamburger	II
Chicken	III

FAVORITE FOOD

Food	Votes
Pizza	7
Macaroni & cheese	6
Spaghetti	6
Frankfurter	5
Chicken	3
Hamburger	2
Steak	1

RETEACH: Interpreting Information

Members of a sixth-grade class were asked to vote on the seven different beverages they would like at a party. The results of the voting are listed in the table at the right. Find the mean, median, mode, and range of this data.

FAVORITE BEVERAGE

Orange juice	3
Apple juice	6
Cranberry juice	6
Lemonade	7
Grape juice	5
Milk	1
Water	2

Remember

Arrange the beverages and scores from the most favorite to the least favorite.

Lemonade	7
Apple juice	6
Cranberry	6
Grape juice	5
Orange juice	3
Water	2
Milk	1
Sum of the scores =	30

To find the **mean,** divide the sum of the scores by the number of scores.

$$\frac{30}{7} = 4.3 \text{ (rounded to nearest tenth)}$$

The **median** is the middle score.
The **mode** is the score that appears most often.
The **range** is the difference between the greatest score and the least score. $7 - 1 = 6$.

The mean is 4.3. The median is 5. The mode is 6. The range is 6.

Copy and complete the table.

	Set	Mean	Median	Mode	Range
1.	5, 6, 7, 5, 12				
2.	10, 4, 10, 6, 10				
3.	3, 2, 6, 2, 2,				
4.	12, 16, 10, 12, 15				

Use with pages 378–379.

RETEACH Making a Pictograph

A survey was taken in Lincoln Elementary School to determine the students' favorite sport. The table below contains the results of this survey.

Use the data from this table to make a pictograph.

FAVORITE SPORT	
Sport	Number of votes
Baseball	40
Basketball	55
Football	35
Volleyball	45
Track	30
Gymnastics	20

Remember

There are six steps for making a pictograph.

1. List the sports on one axis. Use the vertical axis.
2. Choose an appropriate symbol to represent the information on the graph. Choose an *x*.
3. Decide what the symbol will represent. We will let one *x* represent 10 votes.
4. Decide how many symbols are needed for each number.
5. Draw the correct number of symbols on the graph.
6. Write a title on the pictograph. Indicate the symbol we have chosen and its value.

FAVORITE SPORT

Baseball	x x x x
Basketball	x x x x x /
Football	x x x /
Volleyball	x x x x /
Track	x x x
Gymnastics	x x

x = 10 votes
/ = 5 votes

Solve.

1. A survey of the students in Jefferson School revealed the following information.

   Draw a pictograph of this data.

FAVORITE GAMES	
Game	Votes
Kickball	26
Dodgeball	20
Hide and Seek	18
Tag	15

This pictograph was made after a survey at the Washington School. Use it to answer the questions below.

2. What is the pictograph about? _____
3. How many students like jacks? _____
4. How many students like jumprope? _____

FAVORITE GAMES	
Game	Votes
Jacks	x x x /
Jumprope	x x /

x = 10 students   / = 5 students

RETEACH: Making a Bar Graph

This table shows the votes for class president at Kennedy High School. Use this information to make a bar graph.

CLASS PRESIDENT ELECTION

Candidate	Number of votes
John	49
Kay	85
Bob	105
Lisa	70

Remember

There are four steps for making a bar graph.

1. Draw and label the horizontal and vertical axes.
2. Choose appropriate scales and intervals so that the graph will fit the available space.
3. Mark the intervals on the scale. Draw the bars the appropriate length.
4. Write a title on the graph.

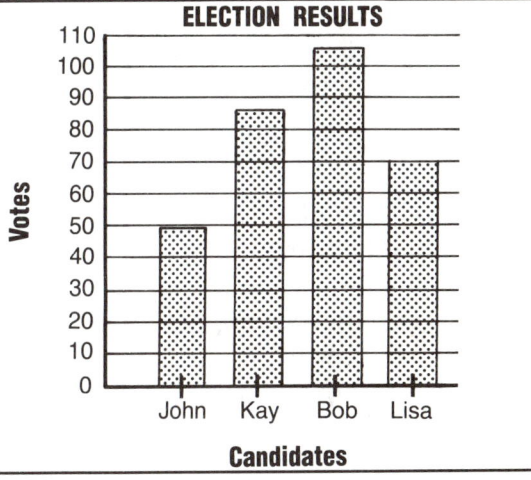

Use the data in this table to make a bar graph.

1. Select five kinds of blue jeans. Then survey 10 people and have them vote for their favorite kind of blue jeans. Tabulate the votes; then make a bar graph of the data.

2. **FAVORITE AUTOMOBILE FOR STUDENTS IN THE WASHINGTON SCHOOL**

Make of automobile	Number of votes
Ford	19
Chevrolet	17
Plymouth	23
Datsun	10
Toyota	15
Continental	16
Volkswagon	4

Use with pages 382–383.

109

RETEACH: Making a Broken-Line Graph

A survey was taken in the Adams School to determine which of six animals the students liked best. The results of the survey are tabulated in the table at the right. Use this information to make a broken-line graph.

FAVORITE ANIMALS

Type of animal	Number of votes
Cat	55
Dog	60
Horse	30
Deer	20
Raccoon	40
Monkey	25

Remember

There are four steps for making a broken-line graph.

1. Study the data. Then draw and label the horizontal and vertical axes.
2. Choose appropriate scales and intervals for the axes.
3. Place a dot on the graph to represent each number in the data.
4. Write a title on the graph.

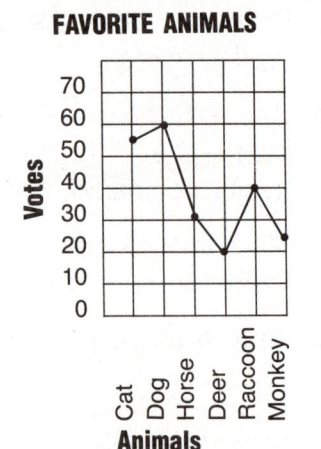

Use the data in this table to make a broken-line graph.

1. FAVORITE MOVIES FOR THE STUDENTS IN MIDWAY SCHOOL

Movies	Number of votes
Space Rockets	45
Melissa	40
Bim Bam	55
Make It Big	50
Beach Talk	60

2. Select five television programs. Then survey 8 students. Have them vote for their favorite TV program. Graph the results on a broken-line graph.

RETEACH: Probability and Expectations

The numbers 1, 2, and 3 are placed in a hat. You draw one number without looking. If you draw an even number, you receive a red bicycle. If you draw an odd number, you receive a blue bicycle. What is the probability that you will receive a red bicycle?

Remember

Probability is the favorable outcomes divided by possible outcomes.

$$\frac{\text{Favorable outcomes}}{\text{Possible outcomes}} \rightarrow \frac{1}{3}$$

The probability of receiving a red bike is $\frac{1}{3}$.

Solve.

A six-sided number cube is tossed. Find the probability of each of the following events.

1. The number showing is a 2. _____

2. The number showing is even. _____

3. The number showing is greater than 3. _____

4. The number showing is less than 8. _____

5. The number showing is prime. _____

6. The number showing is between 2 and 5. _____

7. The number showing is less than 3 or greater than 4.

_____

Use with pages 392–393.

111

RETEACH Independent Events

Mark has 4 sweaters and 3 shirts that can be worn in any combination. How many sweater-shirt combinations does Mark have?

Remember

Multiply to find the number of combinations: the number of first choices times the number of second choices.

Sweater choice	Shirt choices	Choices
4 ×	3 =	12

Mark has 12 sweater-shirt combinations.

Solve.

1. A sailor can send messages by displaying flags of different colors. He has a flag staff that has room for 3 flags. He has 4 flags that he can use for the top position, 3 flags that can be used for the middle position, and 2 flags that can be used for the bottom position. How many combinations of flag messages can be used?

   _____

2. Dustin has 5 kinds of crackers and 3 kinds of cheese. How many cracker-cheese combinations can he make?

   _____

3. Dustin's sister Karen has 7 kinds of crackers and 5 kinds of cheese. How many cracker-cheese combinations can she make? _____

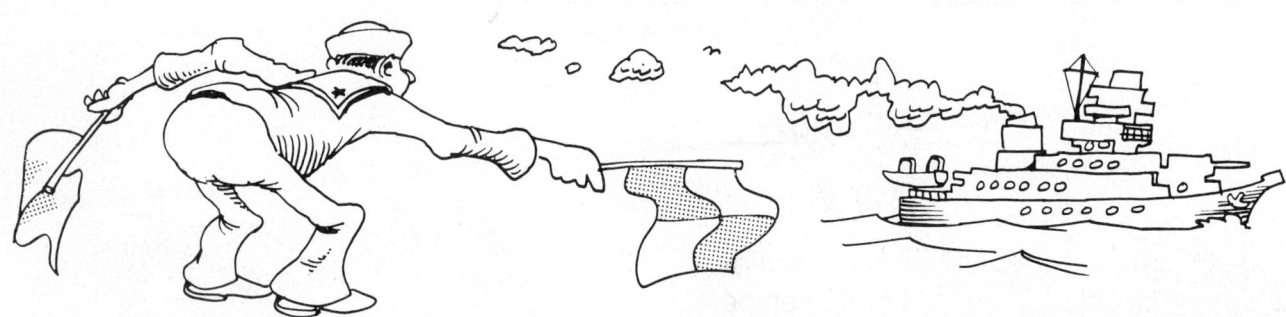

RETEACH   Integers

On cold winter days, the temperature may drop below zero. In Michigan, the temperature can reach 25 below zero. Write that number as an integer.

Remember
Numbers below zero are negative numbers.

25 below zero is ⁻25.

Write an integer to describe each situation.

1. 5° below zero _____
2. a loss of $2 _____
3. 6 steps forward _____
4. a hole 3 ft deep _____
5. score 9 points _____
6. 51 ft above sea level _____
7. 85° above zero _____
8. stock prices drop $6 _____
9. $25 profit _____
10. withdraw $50 _____
11. lose 12 pounds _____
12. 8 steps backward _____

Write the opposite of each situation and the opposite integer.

13. 20 seconds before lift-off _____
14. score 20 points _____
15. up 6 floors _____
16. lose 10 pounds _____
17. reduce the price by $8 _____
18. up 4 flights of stairs _____

Solve.

19. The coldest temperature ever recorded in North Dakota was 60° below zero. Write this temperature as an integer. _____

Use with pages 404–405.

RETEACH Comparing Integers

In many parts of America, people get their water by digging or drilling a hole in the ground. Then they place a pipe in the hole and pump water for the family and for farm animals. The hole and the pipe are called a *well*. Mr. Sydow has a well that is 315 feet deep. Mr. Walker's well is 305 feet deep. Which integer is larger?

Remember

Use a number line to find the larger integer. When two negative numbers are placed on the number line, the number nearer to 0 is the greater.

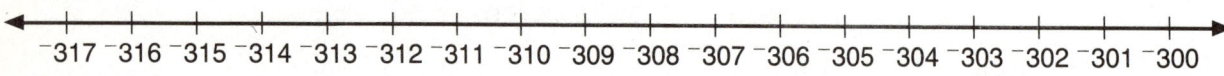

Since ⁻305 is to the right of ⁻315, ⁻305 is larger.

Write >, <, or = for ◯.

1. ⁺6 ◯ ⁺4
2. ⁺3 ◯ ⁺7
3. ⁺5 ◯ ⁺3
4. ⁺6 ◯ ⁺2
5. ⁻5 ◯ ⁻2
6. ⁻7 ◯ ⁻7
7. ⁻2 ◯ ⁻6
8. ⁻9 ◯ ⁻5
9. ⁻3 ◯ ⁺2
10. ⁻6 ◯ ⁺4
11. ⁺10 ◯ ⁻5
12. ⁺4 ◯ ⁻6
13. ⁺12 ◯ ⁺12
14. 0 ◯ ⁻3
15. ⁻8 ◯ 0
16. 0 ◯ 0

Solve.

17. The temperature on Monday was ⁻9°. The temperature on Tuesday was ⁻12°. Which day had the higher temperature?

RETEACH: Adding Integers: Like Signs

One winter day in Mayville, North Dakota, the temperature was ⁻5°. By midnight the temperature had dropped 17°. What was the temperature at midnight?

Remember

The sum of two negative integers is a negative integer. Similarly, the sum of two positive integers is a positive integer.

A number line is a very useful device when you are working problems dealing with integers.

Drop in temperature of 17° is represented by ⁻17°.

(⁻5) + (⁻17) = ⁻22°

The temperature was ⁻22° at midnight.

Add.

1. ⁺3 + ⁺12 = _____
2. ⁻7 + ⁻4 = _____
3. ⁺15 + ⁺8 = _____
4. ⁺12 + ⁺9 = _____
5. ⁺6 + ⁺9 = _____
6. ⁺18 + ⁺9 = _____
7. ⁻20 + ⁻3 = _____
8. ⁻17 + ⁻5 = _____
9. 0 + ⁺5 = _____
10. ⁺16 + ⁺9 = _____
11. ⁻11 + ⁻14 = _____
12. ⁻20 + ⁻13 = _____

Solve.

13. One spring day, the snowdrift by John's house began to melt. The depth of the drift decreased 6 inches. The next day, the depth of the drift decreased 8 inches. Write an integer to describe how the depth of the snowdrift changed during the two days. _____

Use with pages 408–409.

RETEACH Adding Integers: Unlike Signs

When Myrna walked to school one cold winter day, the temperature was ⁻8°. By noon the temperature had climbed 5°. What was the temperature at noon?

Remember

Write the positive and negative signs. Use a number line. ⁻8 + ⁺5 = ◼

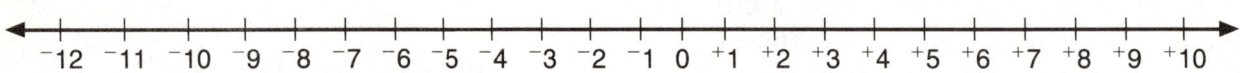

Start at 0. Move left 8 spaces, then move right 5 spaces. The sum has the same sign as the integer farther from zero.

The temperature at noon was ⁻3°.

Add.

1. ⁻6 + ⁺9 = _____
2. ⁻8 + ⁺12 = _____
3. ⁻2 + ⁺5 = _____
4. ⁻5 + ⁺7 = _____
5. ⁻9 + ⁺5 = _____
6. ⁻12 + ⁺8 = _____
7. ⁻8 + ⁺5 = _____
8. ⁻18 + ⁺12 = _____
9. ⁻12 + ⁺7 = _____
10. 0 + ⁻6 = _____
11. ⁻7 + 0 = _____
12. ⁺1 + ⁻9 = _____
13. ⁺7 + ⁻12 = _____
14. ⁺5 + ⁻7 = _____
15. ⁺8 + ⁻10 = _____
16. ⁺7 + ⁻15 = _____
17. ⁻3 + ⁺9 = _____
18. ⁻6 + ⁺5 = _____

Solve.

19. Jimmy went ice-skating on a pond one Saturday morning. The temperature was ⁻12°F. By noon, the temperature had risen 23°F. What was the temperature at noon?

_____

RETEACH: Subtracting Integers

Usually, the temperature outside rises during the day and falls at night. During one 24-hour period, the temperature rose 9° and fell 15°. How much did the temperature change?

Remember

To subtract an integer, add its opposite.
A rise of 9° and a fall of 15° could be written as

$$^+9 + {}^-15.$$

If you think of a fall in temperature as subtracting, you write the problem this way

$$^+9 - {}^+15.$$

Subtracting 15 is the same as adding $^-15$. Hence,
$^+9 - {}^+15 = {}^+9 + {}^-15 = {}^-6$.

The temperature dropped 6° during the period.

Subtract.

1. $^+17 - {}^+11 =$ _____
2. $^+6 - {}^+8 =$ _____
3. $^+5 - {}^+15 =$ _____
4. $^-13 - {}^-6 =$ _____
5. $^-4 - {}^-10 =$ _____
6. $0 - {}^-5 =$ _____
7. $^+6 - {}^+13 =$ _____
8. $^+3 - {}^-12 =$ _____
9. $^-18 - {}^+13 =$ _____
10. $^-5 - {}^-4 =$ _____
11. $^-15 - {}^+3 =$ _____
12. $^-5 - {}^-4 =$ _____

Solve.

13. One morning, the temperature was 22°. During the morning, the temperature rose 7°. During the afternoon, the temperature dropped 15°. What was the temperature by evening? _____

Use with pages 412–413.

RETEACH Graphing Ordered Pairs

Many years ago, a French mathematician discovered he could label the position of a fly on the ceiling by using just two integers. He called the two integers an **ordered pair.** He used a coordinate grid to divide the ceiling into four sections.

Use an ordered pair to label point A on the grid.

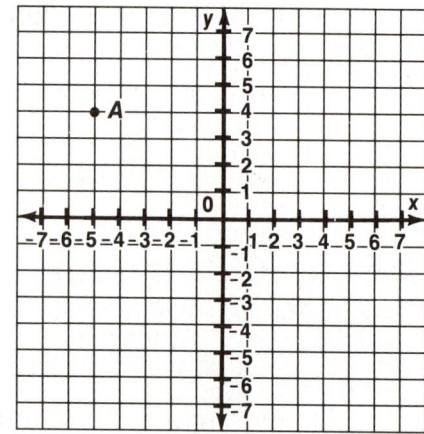

Remember

To locate point A, we must move 5 units left and 4 units up. The first number of the ordered pair shows how to move across.

Point A is located at (⁻5,4).

Solve.

Captain Kidd buried some treasure. He left a map and a set of directions.

1. The map is a coordinate grid with its center at an old oak tree. The directions say the treasure was buried at various places. Treasure was buried at (5,4), (⁻3,2), (4,⁻2), (⁻3,⁻3), (⁻5,4), (3,⁻4), (⁻2,⁻2), and (4,1). Use graph paper to make a grid like the one on the right. Locate all of the treasure on the map.

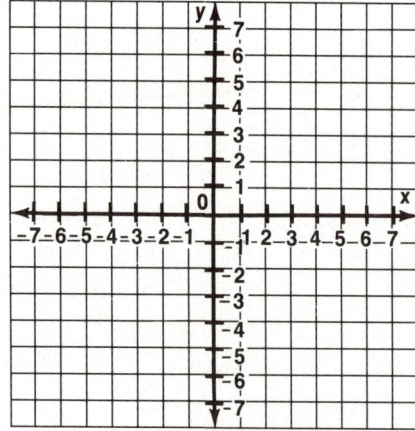

2. Captain Kidd also left a map of his camp. Write an ordered pair to describe the location of the following:

campfire _____   tents _____

stream _____   oak tree _____

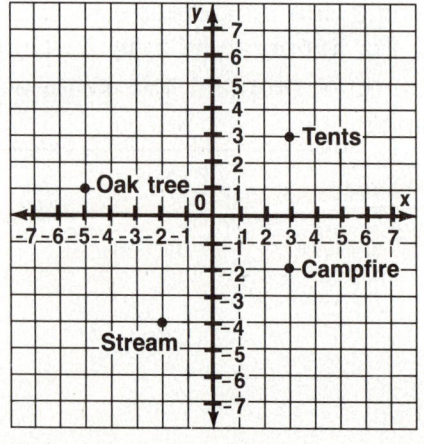